Preface

Although the current syllabus for Higher grade Biology was introduced several years ago (candidates were first presented for the 'new' Higher grade examination in 1971), no book has been published which satisfactorily covers the course, prior to this text.

This book offers, for the first time, a complete collection of student notes adhering strictly to the 'H' grade Biology syllabus (and assuming knowledge of SCEEB Ordinary grade Biology). It presents the essential facts and ideas required for this level in the most concise and visual manner possible. All areas of the course are covered without the inclusion of irrelevant extras. The book can, therefore, be employed as:

(a) a tool of reinforcement and consolidation throughout the class teaching of the entire 'H' grade course. In this respect it is particularly suitable for both the average and the borderline candidate.

(b) an excellent substitute for the notes which many teachers issue, out of necessity, to Higher pupils at present. Chapters are deliberately short to allow for maximum flexibility in teaching order.

(c) a comprehensive revision text of all the essential facts. In this respect it is suitable for use by all Higher Biology candidates on completion of the course, especially since each chapter ends with several revision questions intended to reinforce and test the understanding of fundamental ideas.

JT *et al.*

Contents

Higher Biology

Team Co-ordinator
James Torrance

Writing Team
James Torrance
James Fullarton
Clare Marsh
Robert Rowatt
James Simms
Caroline Stevenson

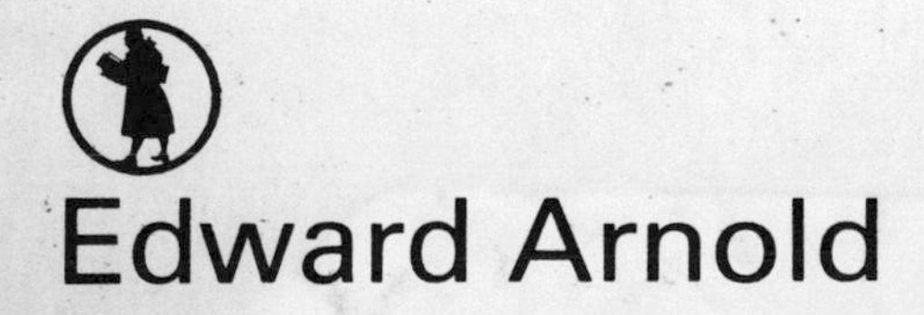

Edward Arnold

First published in Great Britain 1979
by Edward Arnold (Publishers) Ltd
41 Bedford Square, London WC1B 3DQ

Edward Arnold (Australia) Pty Ltd
80 Waverley Road
Caulfield East
Victoria 3145, Australia

Reprinted 1980, 1981, 1982, 1984, 1985

British Library Cataloguing in Publication Data

ISBN 0 7131 0336 1

Diagrams by James Torrance

Printed and bound in Hong Kong
by Wing King Tong Co., Ltd.

1 Ultra-structure of the cell

The maximum magnification afforded by a light microscope is 1500 times. An **electron microscope** can achieve magnifications of over 500 000 times. However material to be viewed in an electron microscope must be fixed, stained, sliced and mounted in a vacuum. As a result, **artifacts** may occur since the material being viewed is dead and is possibly also damaged or distorted.

Units of measurement

1 metre (m) = 1000 millimetres (mm)
1 millimetre (mm) = 1000 micrometres (microns) (μm)
1 micrometre (μm) = 1000 nanometres (nm)

Ultra-structures

The **cell wall** is a non living layer of cellulose fibres. It is freely permeable to aqueous solutions and maintains the shape of the plant cell.
The **cell membrane** (plasmalemma) is a single **unit membrane** (figure 1.2) which is selectively permeable and involved in osmosis and active uptake of ions and molecules.
The **vacuole** contains cell sap and in conjunction with the cytoplasm determines the cell's osmotic potential.
The **cytoplasm** is a fluid material in which the organelles are suspended.
The **nuclear membrane** is a double unit membrane continuous with the endoplasmic reticulum.
Pores in the nuclear membrane allow exchanges of materials between nucleus and cytoplasm.
The **nucleus** controls the cell and contains chromosomes which are visible during cell division. It is the site of RNA synthesis.
The **nucleolus** is the site of one type of RNA.

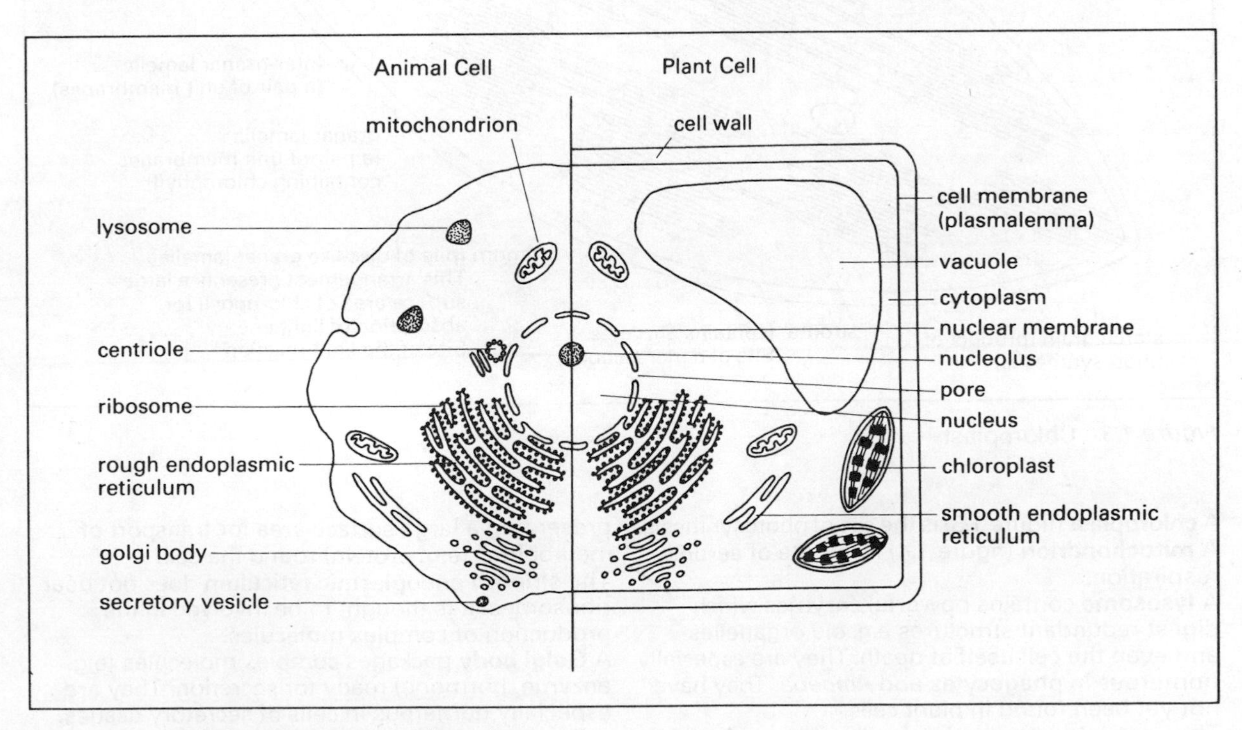

Figure 1.1 Cellular ultra-structure based on several electron micrographs (left side = typical animal cell, right side = typical plant cell). Magnification = X10 000 approximately

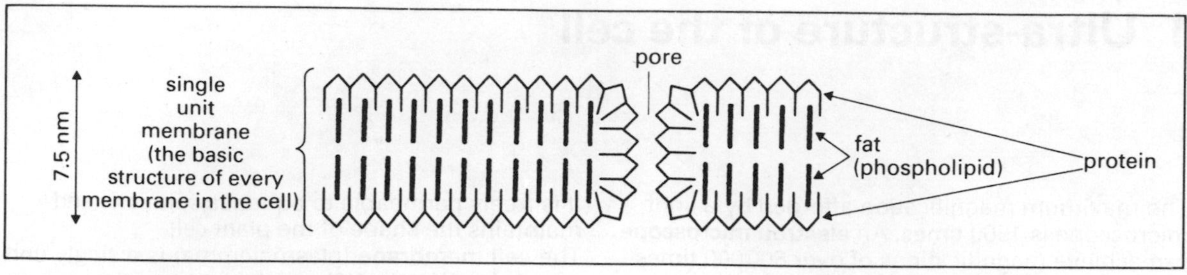

Figure 1.2 Structure of a unit membrane

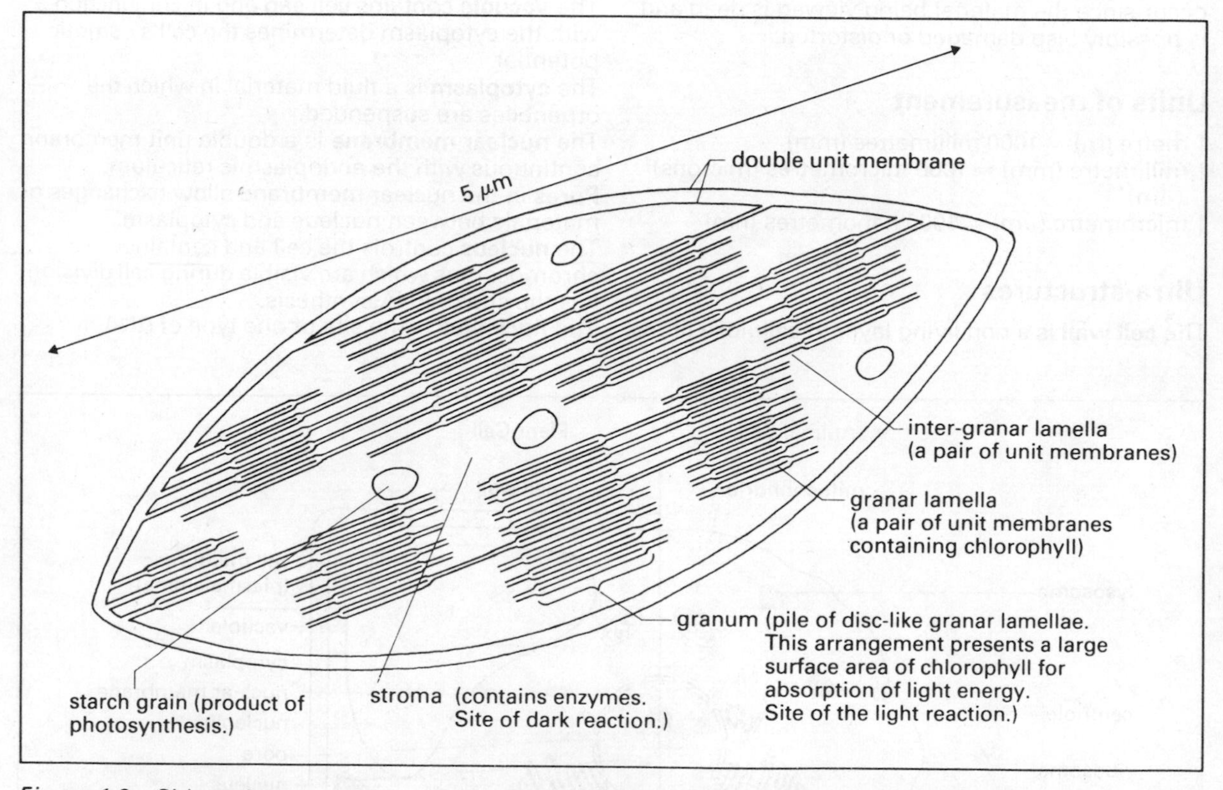

Figure 1.3 Chloroplast

A **chloroplast** (figure 1.3) is the site of photosynthesis.
A **mitochondrion** (figure 1.4) is the site of aerobic respiration.
A **lysosome** contains powerful enzymes which digest redundant structures e.g. old organelles and even the cell itself at death. They are especially numerous in phagocytes and *Amoeba*. They have not yet been found in plant cells.
The **centrioles** are involved in the formation of spindle fibres during cell division.
A **ribosome** is the site of protein synthesis.
The **rough endoplasmic reticulum** is a system of membranes bearing ribosomes on the outside and presenting a large surface area for transport of metabolites (e.g. protein) round the cell.
The **smooth endoplasmic reticulum** does not bear ribosomes. It is thought to be involved in the production of complex molecules.
A **Golgi body** packages complex molecules (e.g. enzyme, hormone) ready for secretion. They are especially numerous in cells of secretory tissues.

Each organelle is structurally adapted to perform a particular function. The subcellular organisation of a cell is, therefore, in many ways analogous to the organisation found amongst the organs of a complex multicellular organism.

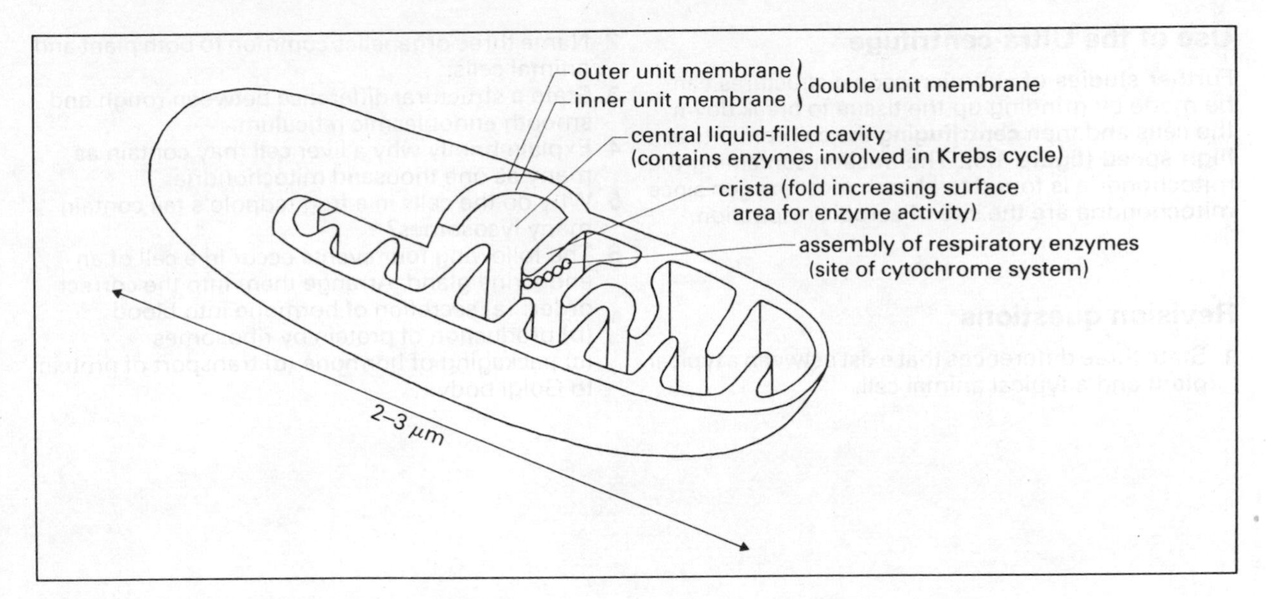

Figure 1.4 Mitochondrion

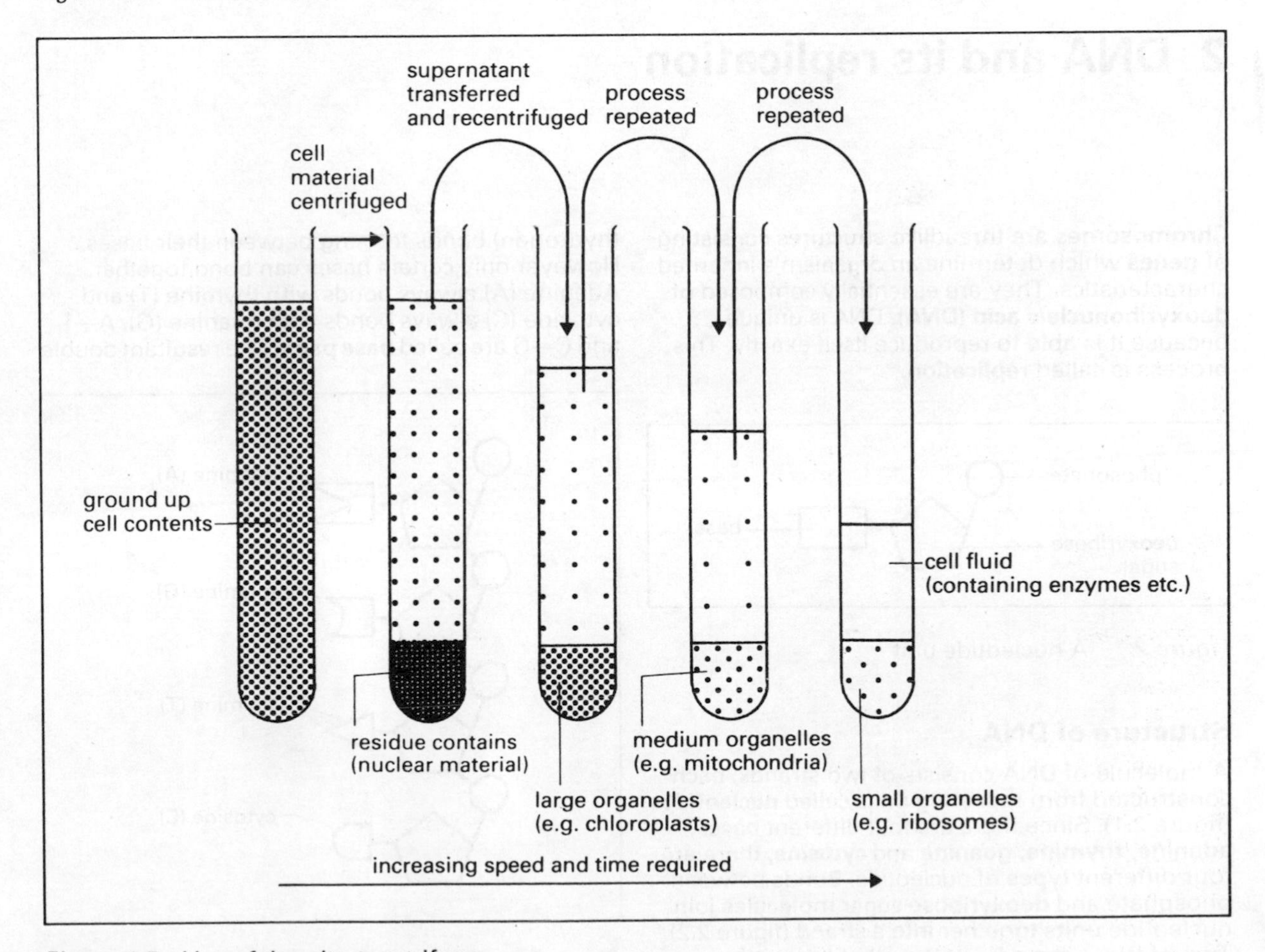

Figure 1.5 Use of the ultra-centrifuge

Use of the Ultra-centrifuge

Further studies of submicroscopic structures can be made by grinding up the tissue to break down the cells and then **centrifuging** the cell contents at high speed (figure 1.5). The fraction containing the mitochondria is found to absorb most oxygen since mitochondria are the site of aerobic respiration.

Revision questions

1 State three differences that exist between a typical plant and a typical animal cell.

2 Name three organelles common to both plant and animal cells.

3 State a structural difference between rough and smooth endoplasmic reticulum.

4 Explain briefly why a liver cell may contain as many as one thousand mitochondria.

5 Why do the cells in a frog tadpole's tail contain many lysosomes?

6 The following four events occur in a cell of an endocrine gland. Arrange them into the correct order: (a) secretion of hormone into blood (b) production of protein by ribosomes (c) packaging of hormone (d) transport of protein to Golgi body.

2 DNA and its replication

Chromosomes are threadlike structures consisting of **genes** which determine an organism's inherited characteristics. They are essentially composed of **deoxyribonucleic acid (DNA).** DNA is unique because it is able to reproduce itself exactly. This process is called **replication.**

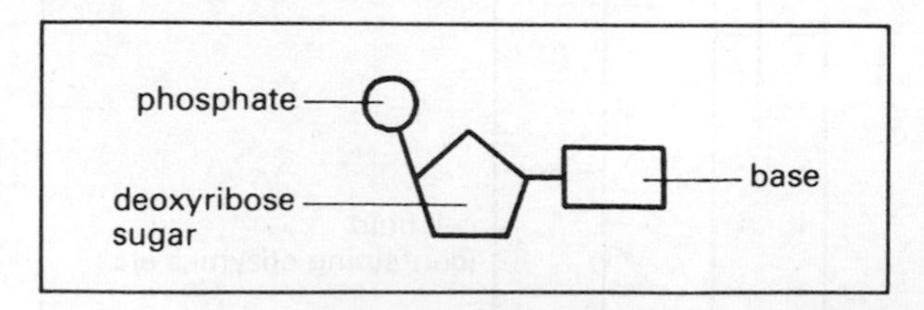

Figure 2.1 A nucleotide unit

Structure of DNA

A molecule of DNA consists of two strands, each constructed from repeating units called **nucleotides** (figure 2.1). Since there are four different bases, **adenine, thymine, guanine** and **cytosine,** there are four different types of nucleotide. Bonds between phosphate and **deoxyribose** sugar molecules join nucleotide units together into a strand (figure 2.2). Two of these strands join together by weak (hydrogen) bonds forming between their bases. However only certain bases can bond together. Adenine (A) always bonds with thymine (T) and cytosine (C) always bonds with guanine (G). A—T and C—G are called **base pairs.** The resultant double

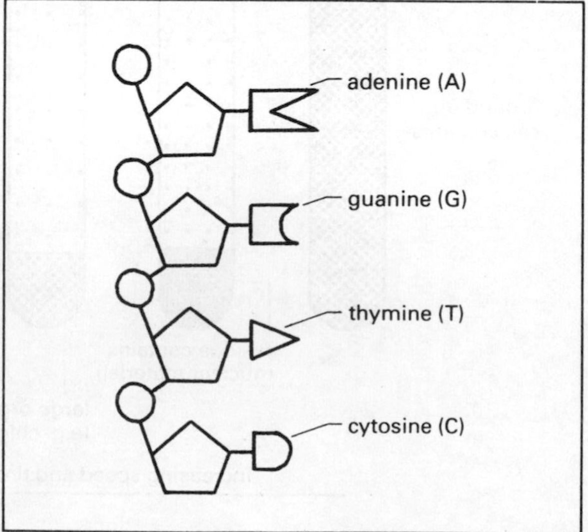

Figure 2.2 Sugar-phosphate 'backbone' structure

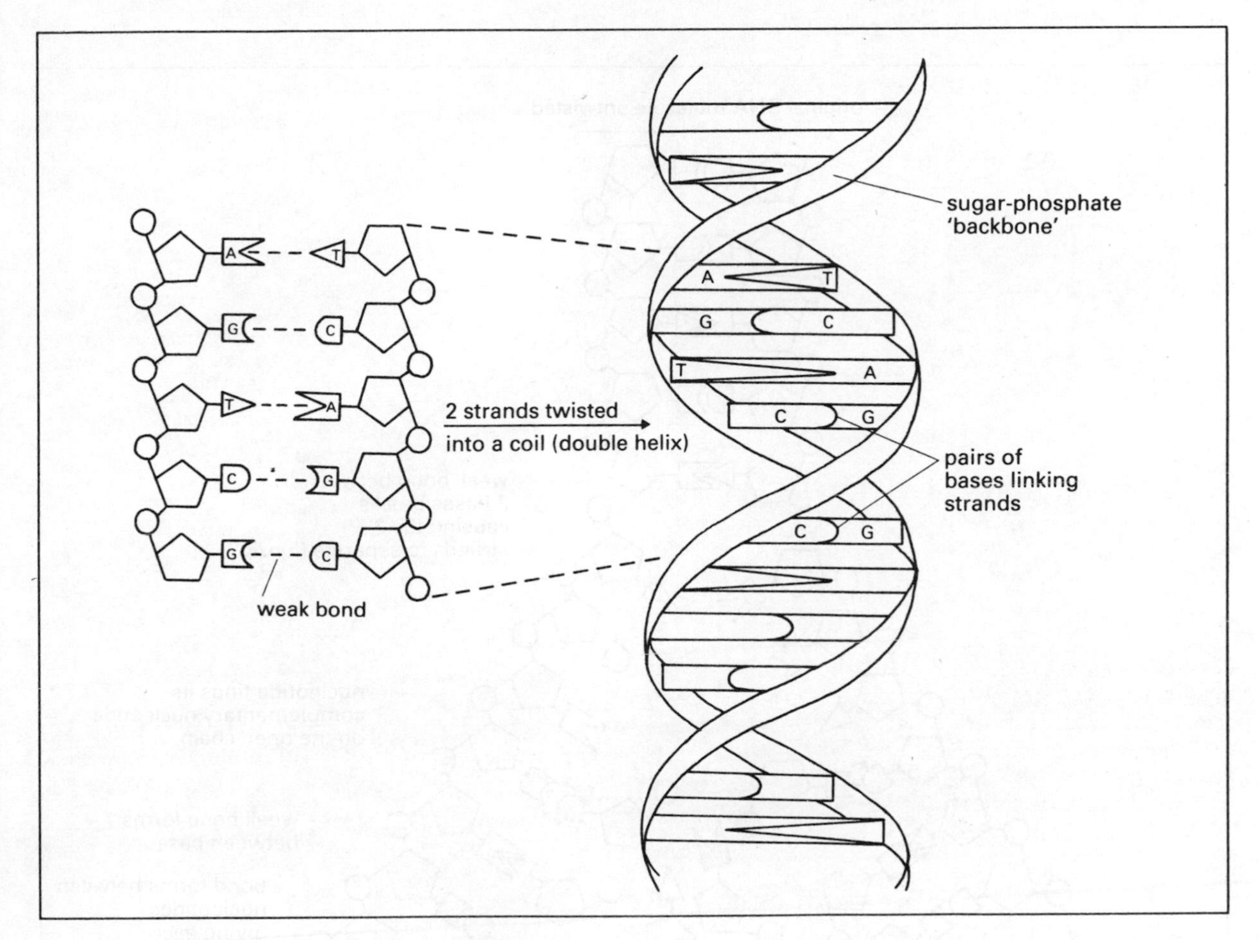

Figure 2.3 Structure of DNA

stranded molecule is DNA and in order to obey the base pairing rule it has its two strands arranged as shown in figure 2.3. The structure is like a coiled ladder in which the sugar and phosphate groups are the 'uprights' and the base pairs the 'rungs'.

Replication of DNA

If a supply of DNA (to act as a template for the new molecules), each of the four nucleotides, the appropriate enzymes and ATP (for energy) are all present, then DNA replication can occur as shown in figure 2.4. The process continues until the whole DNA molecule has replicated. This results in the formation of two new molecules each of which has exactly the same sequence of bases as the parent molecule. As each of the two new molecules forms, it coils itself into a **double helix** (figure 2.6). The whole process occurs in the nucleus of the cell. The structure and replication of DNA was first described by Watson and Crick in 1953.

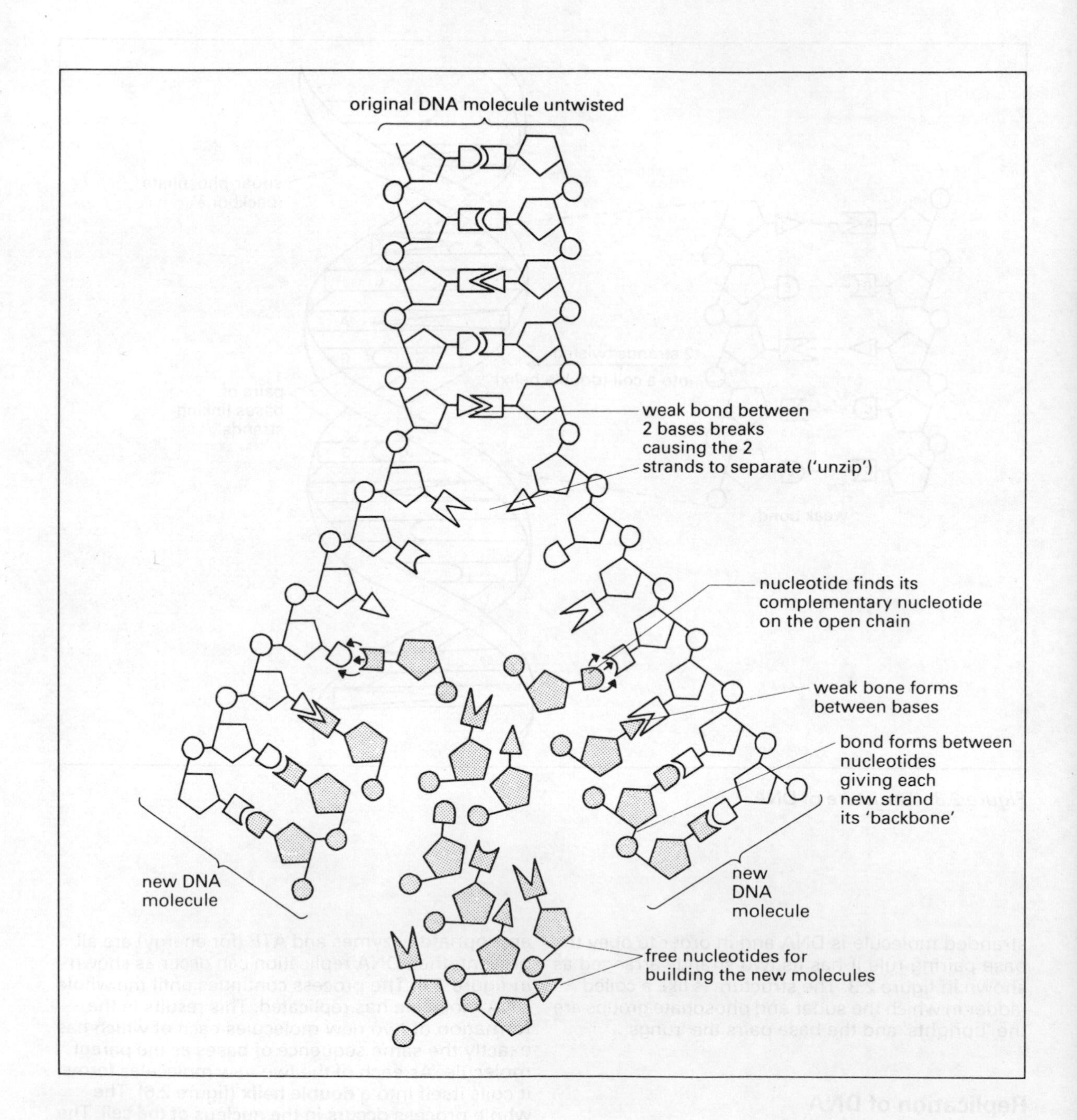

Figure 2.4 Replication of DNA

Revision questions

1 Figure 2.5 shows part of a single DNA strand.
(a) Of which molecules is region X composed?
(b) Draw the strand complementary to the one shown.

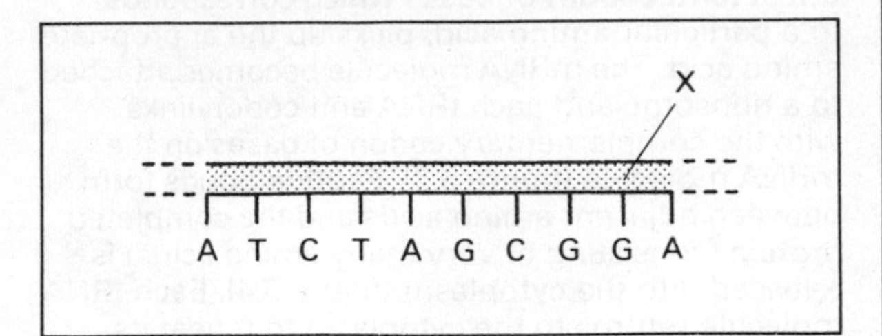

Figure 2.5

2 Draw a diagram to show two complementary nucleotides in opposite strands of a DNA molecule.
3 What proportion of a parent DNA molecule is contained by each daughter molecule after replication?
4 What substances must be present before DNA replication can take place?
5 Place the following events that occur during DNA replication, in the correct order: (a) bonds between opposite bases break (b) bonds between opposite bases form (c) DNA molecule uncoils from one end (d) opposite strands separate (e) daughter molecules coil into double helices (f) sugar-phosphate bonds form (g) free nucleotides find their opposite nucleotide on each strand.

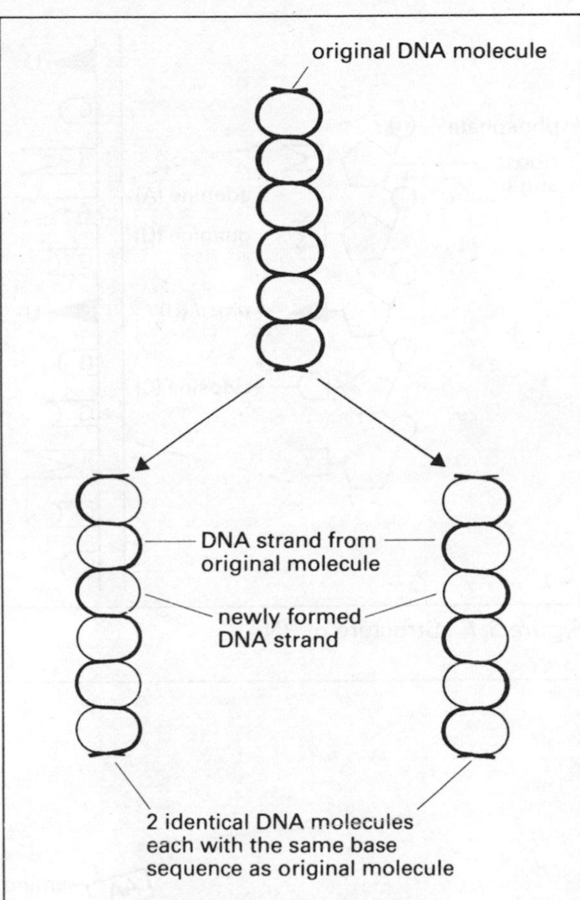

Figure 2.6 Summary of DNA replication

3 RNA and protein synthesis

The information for the construction of a protein with the correct sequence of amino acids is carried from the gene by a template molecule called **ribonucleic acid (RNA)** (figure 3.1). RNA is a large molecule similar in structure to a single strand of DNA but containing the base **uracil** in place of thymine and **ribose** sugar in place of deoxyribose. Two types of RNA are **messenger RNA (mRNA)** and **transfer RNA (tRNA)**.

Protein synthesis

Part of a DNA molecule in a gene splits open exposing its bases (figure 3.2). A molecule of mRNA

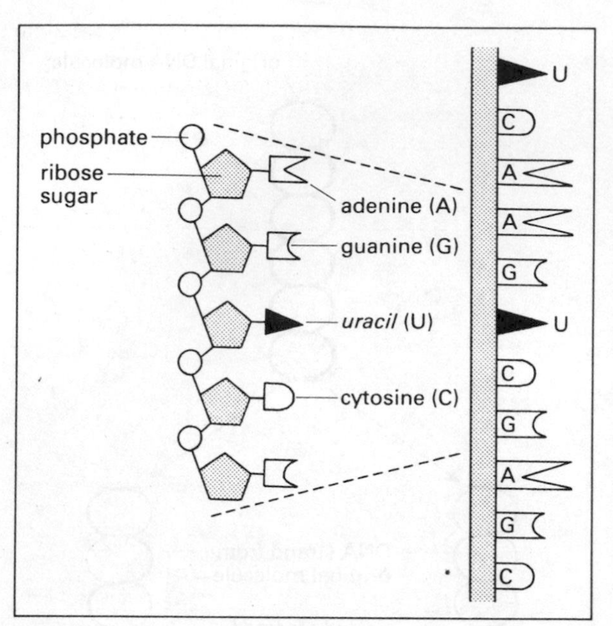

Figure 3.1 Structure of RNA

is transcribed from one DNA strand using free nucleotides. Complementary base pairing of A—U and C—G determines the sequence of the bases in the mRNA molecule. This mRNA template leaves the nucleus and enters the cytoplasm. Meanwhile in the cytoplasm, each tRNA molecule, bearing a triplet **(anti-codon)** of bases which corresponds to a particular **amino acid,** picks up the appropriate amino acid. The mRNA molecule becomes attached to a ribosome and each tRNA anti-codon links with the complementary **codon** of bases on the mRNA molecule (figure 3.3). **Peptide** bonds form between adjacent amino acids and the completed **protein** (consisting of very many amino acids) is released into the cytoplasm (figure 3.4). Each tRNA molecule returns to the cytoplasm to repeat its function and the mRNA molecule may be reused to produce another molecule of the protein.

Proteins may be used within the cell (e.g. as enzymes) or may be exported for use in other cells of the organism (e.g. for growth and repair). The properties and functions of all proteins are determined by the sequence of their component

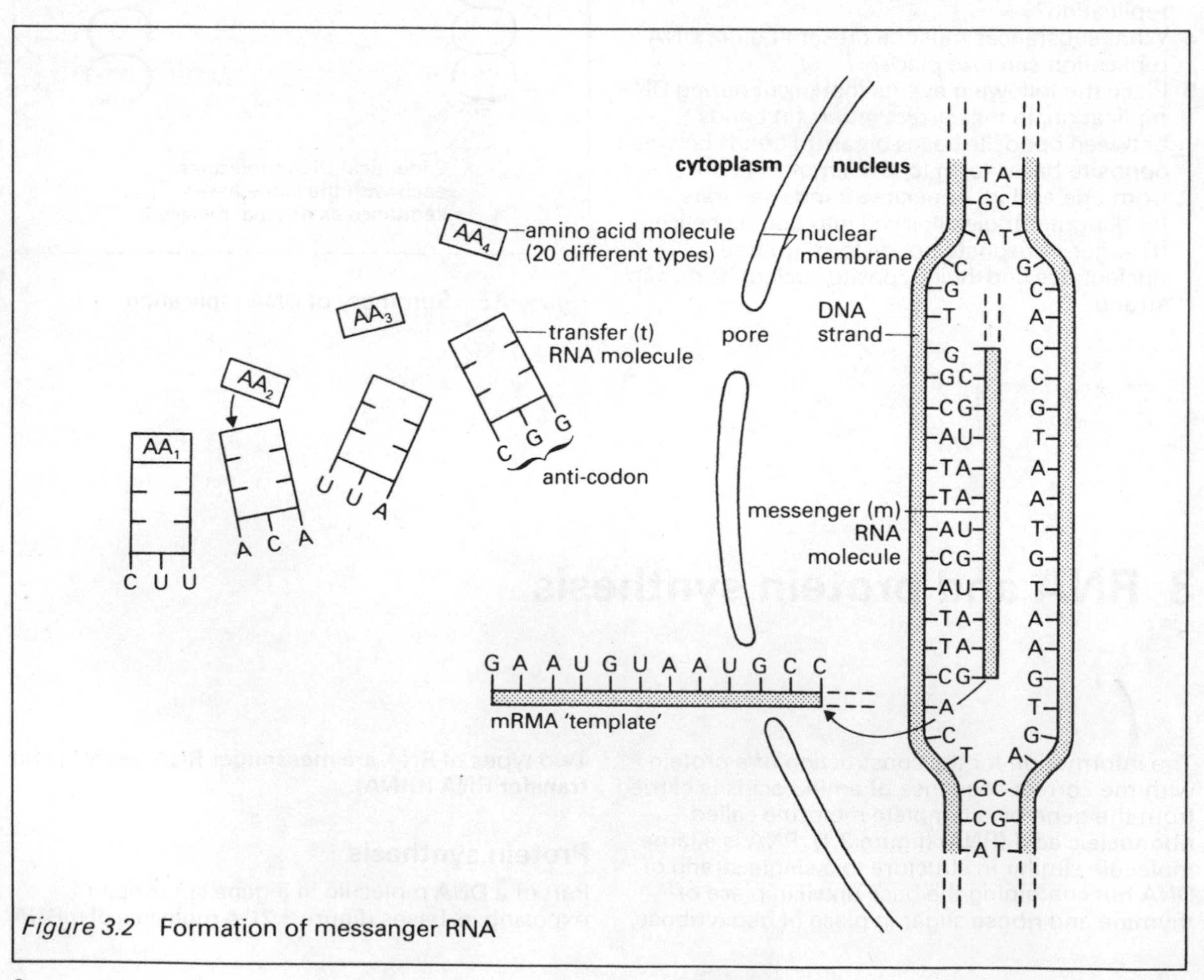

Figure 3.2 Formation of messenger RNA

amino acids. This sequence is, in turn, determined by the original sequence of the bases present in the part of the DNA molecule (the gene) that codes for the protein.

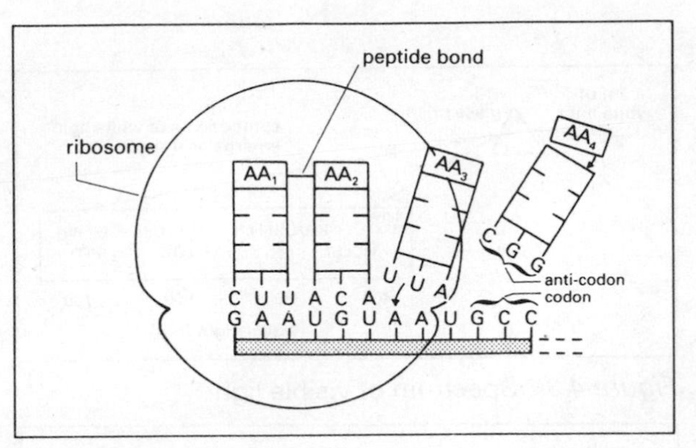

Figure 3.3 Attachment of mRNA to a ribosome

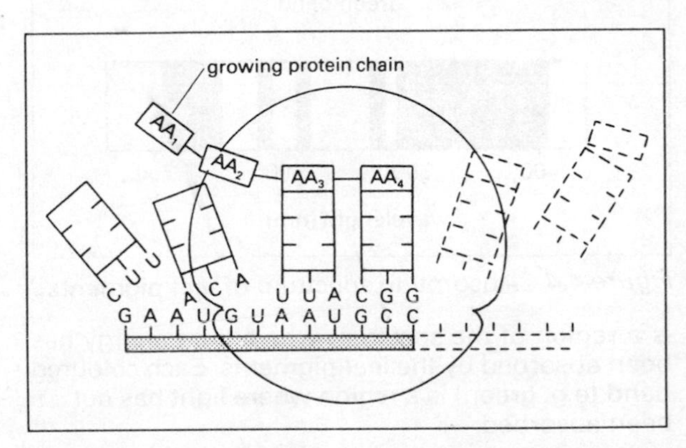

Figure 3.4 Translation of mRNA into protein

Revision questions

1 Draw a diagram of the RNA strand that would be transcribed from section X of the DNA molecule shown in figure 3.5.

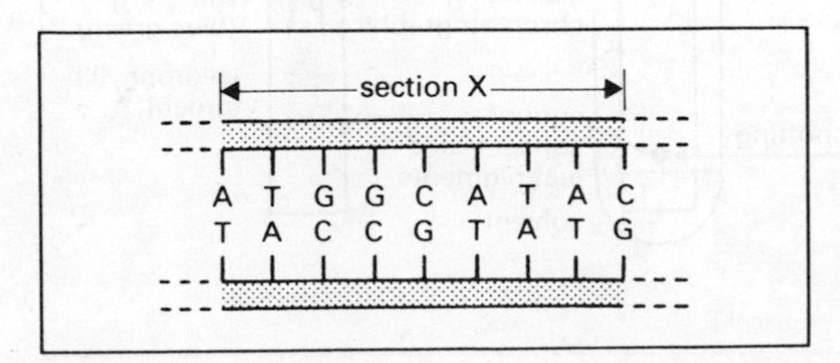

Figure 3.5

2 Using the information given in table 3.1, (a) write down the base codon (mRNA) for each amino acid, (b) draw a diagram of the sequence of amino acids that would be formed from the portion of mRNA shown in figure 3.6.

Amino acid	Base anti-codon (tRNA)
asparagine	UUA
glutamic acid	CUU
proline	GGA
threonine	UGG
tyrosine	AUA

Table 3.1

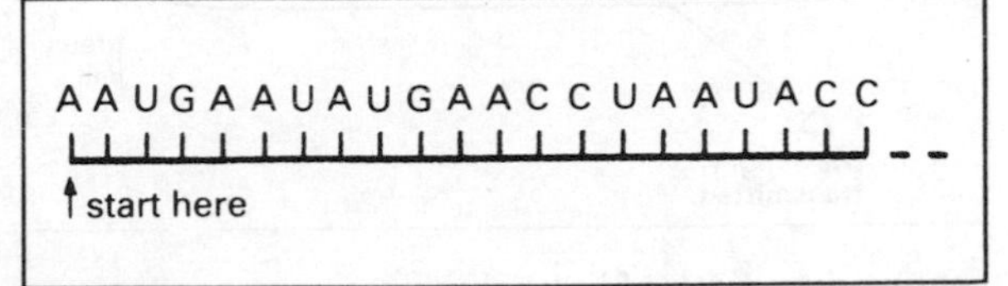

Figure 3.6

3 Copy and complete the following:

stage of synthesis	**site**
formation of mRNA	
tRNA collects amino acid	
formation of codon-anticodon base links	

4 What finally happens to (a) the tRNA molecule (b) the mRNA molecule (c) the completed protein?

4 Absorption of light by leaf pigments

Although most of the light striking a leaf is absorbed, only a small part of it is used for photosynthesis (figure 4.1). This light energy is trapped by the **leaf pigments** which can be extracted and separated as shown in figure 4.2.

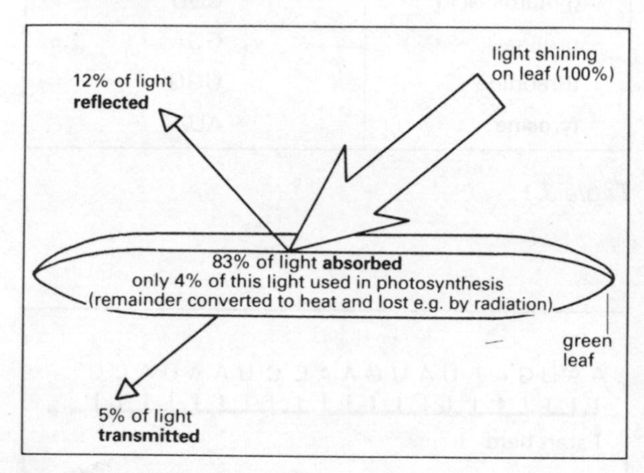

Figure 4.1 Fate of light shining on a green leaf

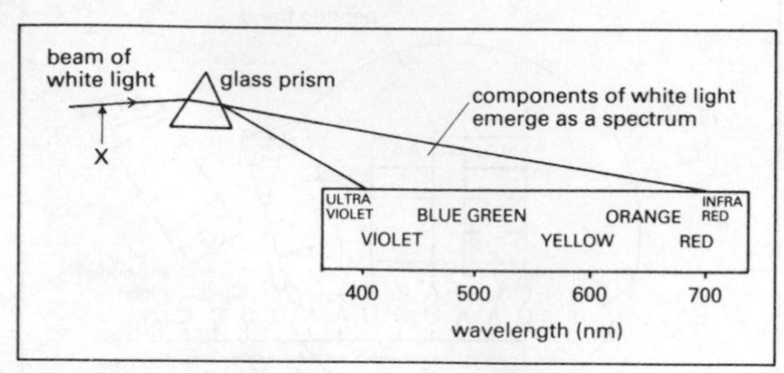

Figure 4.3 Spectrum of visible light

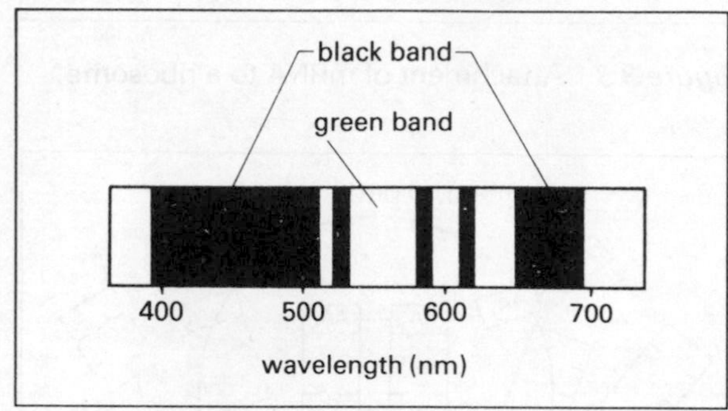

Figure 4.4 Absorption spectrum of leaf pigments

Absorption spectra of leaf pigments

When a beam of white light is first passed through a sample of leaf pigments (at point X, figure 4.3) and then through a glass prism, an **absorption spectrum** (figure 4.4) is produced. Each black band is a region of the spectrum where light energy has been absorbed by the leaf pigments. Each coloured band (e.g. green) is a region where light has not been absorbed.

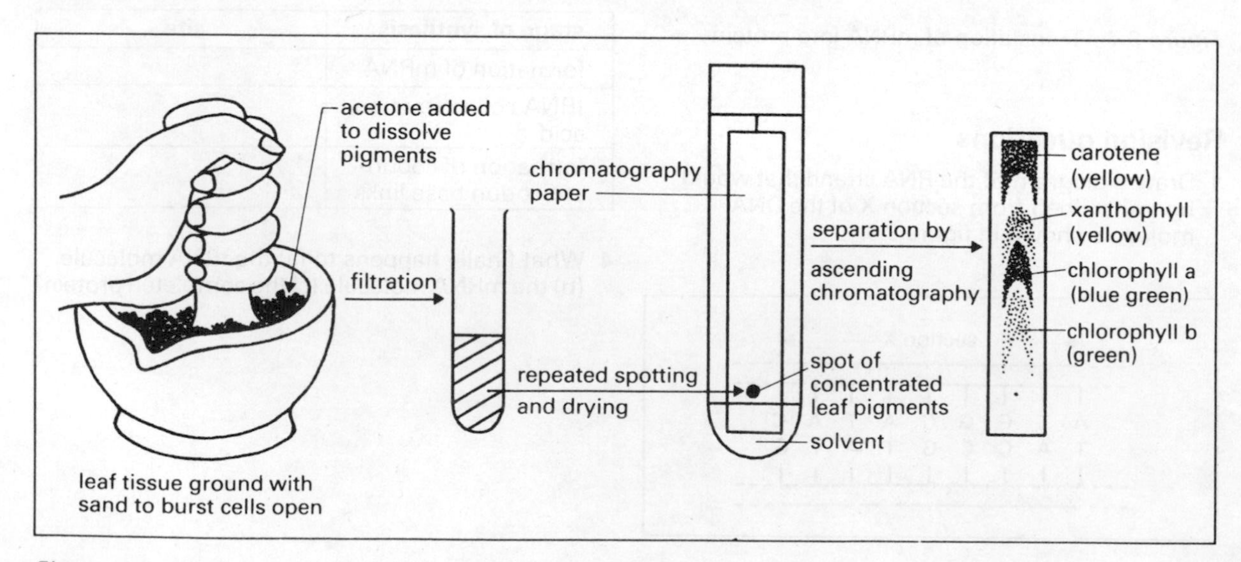

Figure 4.2 Extraction and separation of leaf pigments

A **spectrometer** is used to measure the amount of light absorbed by the leaf pigments at each wavelength of light. The information obtained can be graphed (figure 4.5).

The wavelengths of light not absorbed by a pigment are transmitted or reflected. Hence chlorophyll is green and carotene and xanthophall are yellow

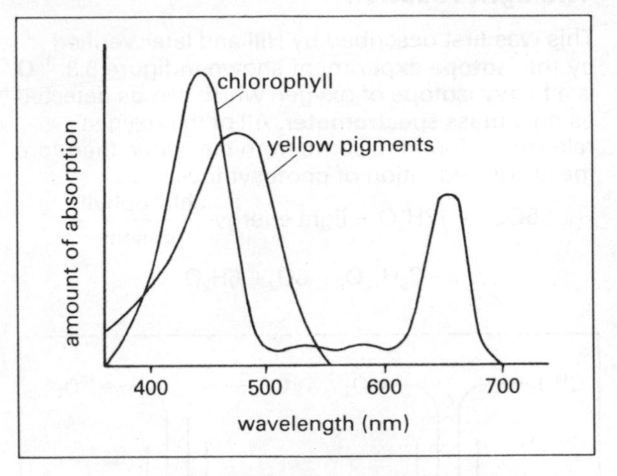

Figure 4.5 Graph of absorption spectrum of leaf pigments

Fluorescence

When a chlorophyll molecule absorbs light, some of its electrons become 'excited' and raised to a higher energy level. In a solution of chlorophyll, the excited electrons quickly drop back to the original energy level giving out the energy as red light (**fluorescence**). In living cells, the energy is used for photosynthesis and therefore is not fluoresced.

Revision questions

1 What are the possible fates of light energy striking a green leaf?
2 Name four pigments present in a green leaf.
3 What technique is used to separate these pigments?
4 Why is chlorophyll a, blue-green in colour?
5 (a) Which pigments absorb most light enery at region Z in the graph shown in figure 4.6?
(b) Is this energy used in photosynthesis? Explain your answer.

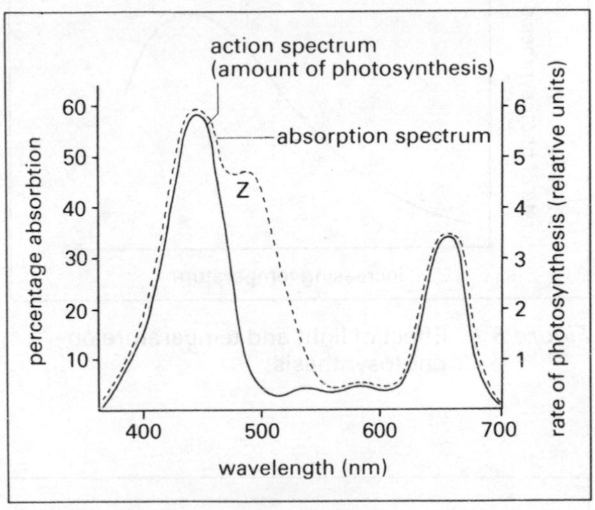

Figure 4.6

5 Chemistry of photosynthesis

Equation: $6CO_2 + 6H_2O +$ light energy $\xrightarrow[\text{present}]{\text{chlorophyll}}$

$$C_6H_{12}O_6 + 6O_2$$

Evidence that photosynthesis consists of two separate reactions

When temperature is kept constant, increasing light intensity has the effect shown in figure 5.1. This proves that part of the reaction is **light dependent (photochemical)**. When light intensity is kept constant but temperature varied, then the effect shown in the second graph results, proving that part of photosynthesis is **temperature dependent (thermochemical)**. Therefore photosynthesis must consist of a reaction dependent on light—the **light reaction** and a reaction dependent on temperature—the **dark reaction.**

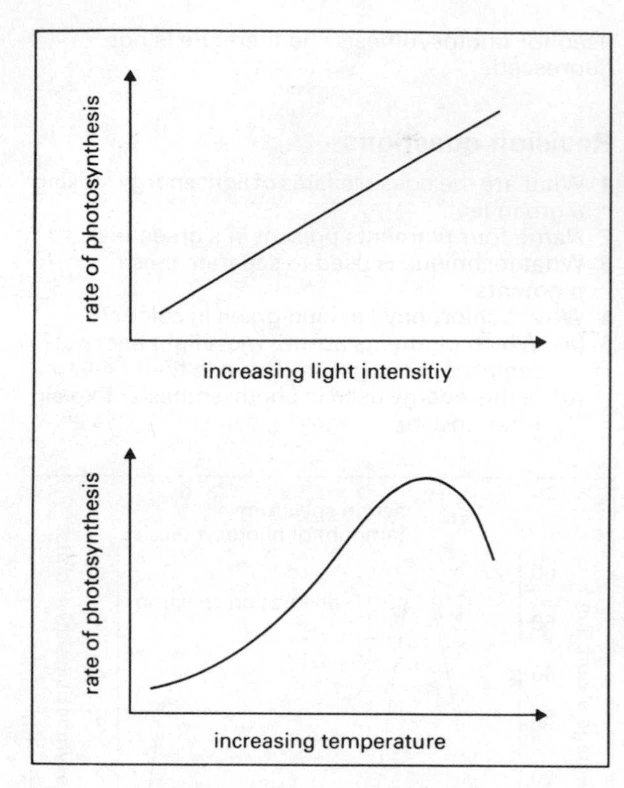

Figure 5.1 Effect of light and temperature on photosynthesis

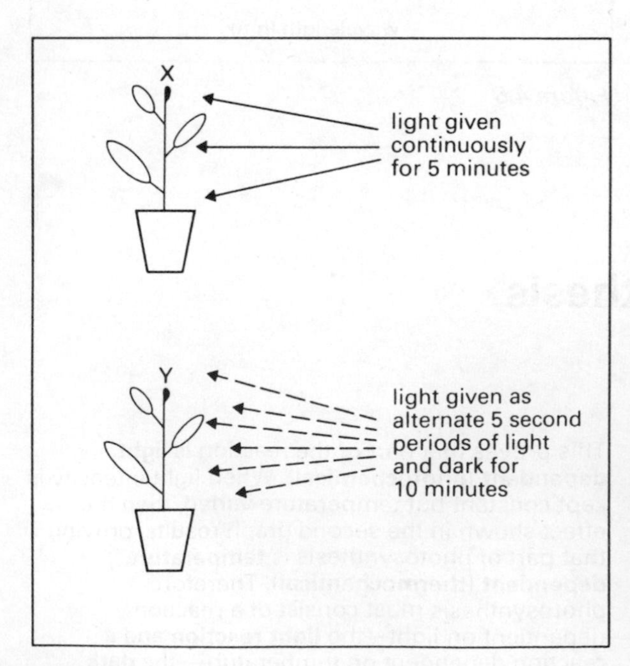

Figure 5.2 Effect of continous and intermittent light

In the experiment shown in figure 5.2, each plant receives the same total amount of light, yet plant Y produces larger amounts of photosynthetic products than plant X. This suggests that the light reaction quickly produces metabolites which are slowly used up during the dark reaction.

The light reaction

This was first described by Hill and later verified by the **isotope** experiment shown in figure 5.3. ^{18}**O** is a **heavy** isotope of oxygen which can be detected using a **mass spectrometer.** All of the oxygen released is found to come from the water, therefore the correct equation of photosynthesis is

$$6CO_2 + 12H_2O + \text{light energy} \xrightarrow[\text{present}]{\text{chlorophyll}} C_6H_{12}O_6 + 6O_2 + 6H_2O$$

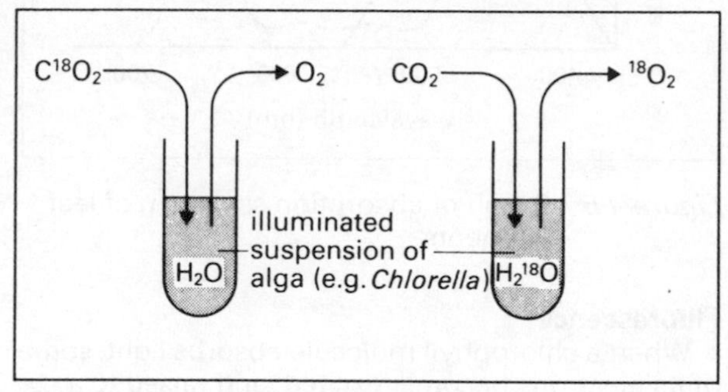

Figure 5.3 Isotope experiment

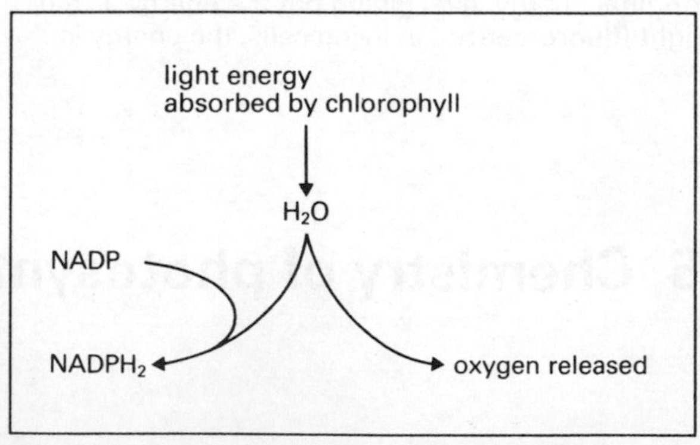

Figure 5.4 The light reaction (Hill's reaction)

Thus during the light reaction (figure 5.4) **photolysis** of water occurs. The **hydrogen acceptor** involved is called **nicotinamide adenine dinucleotide phosphate (NADP).** The light reaction, which occurs in the **granum** of a chloroplast, also converts light energy to chemical energy held in ATP as shown in figures 5.5 and 6.

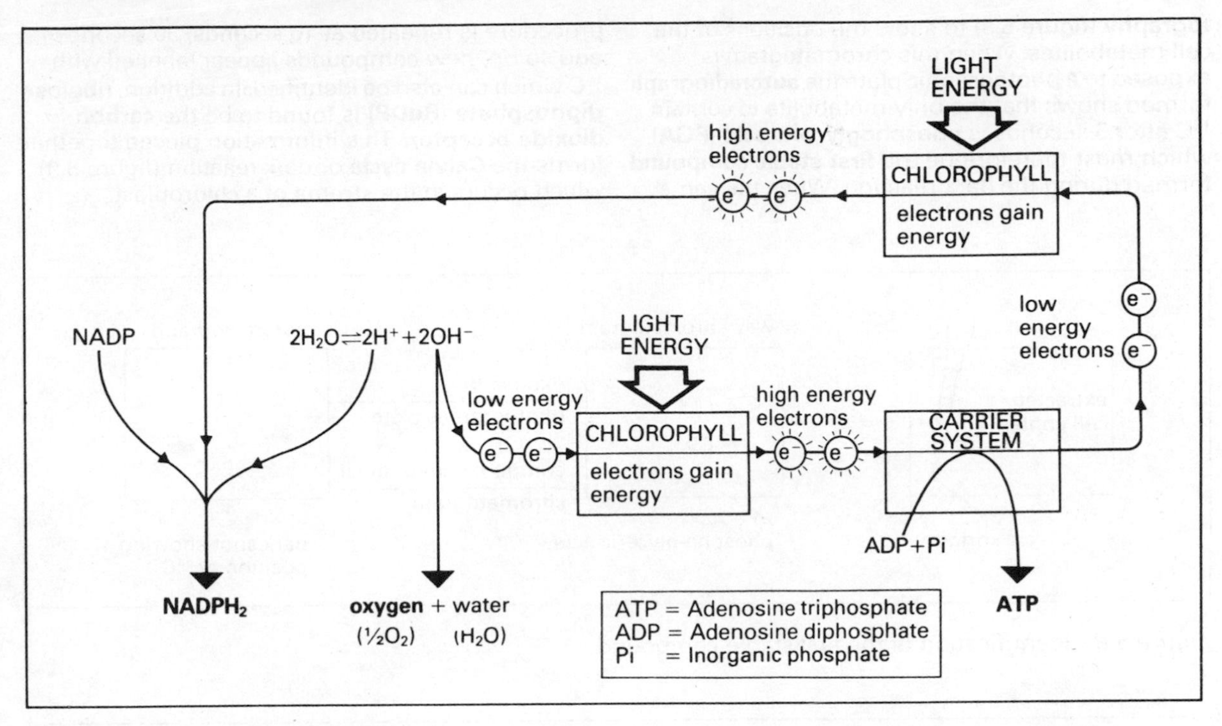

Figure 5.5 Non-cyclic photophosphorylation

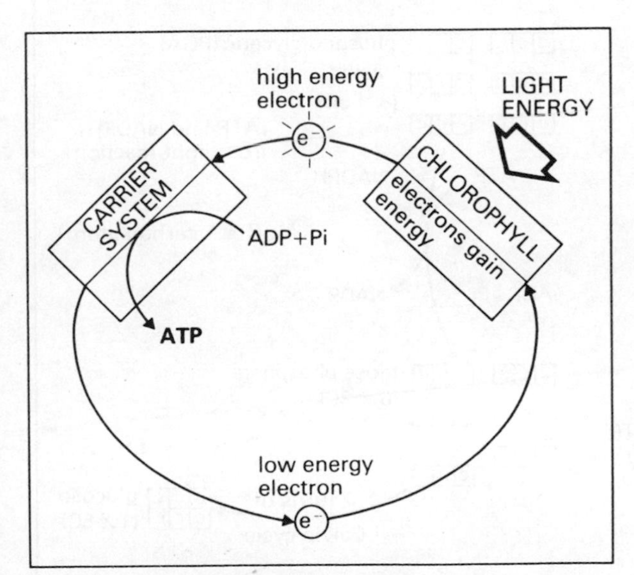

Figure 5.6 Cyclic photophosphorylation

The dark reaction

This was discovered by Calvin using the apparatus shown in figure 5.7. ^{14}C is a **radioactive** isotope of carbon. The unicellular alga *Chlorella* is a suitable plant to use because alcohol quickly kills the cells which then allow their contents to leak out.

The shape of the flask exposes a large surface area of plant material to light. The sampling valve is opened to release the first sample of cells after 5 seconds of exposure to $^{14}CO_2$. The cell contents are spotted and then separated by two-way **chroma-**

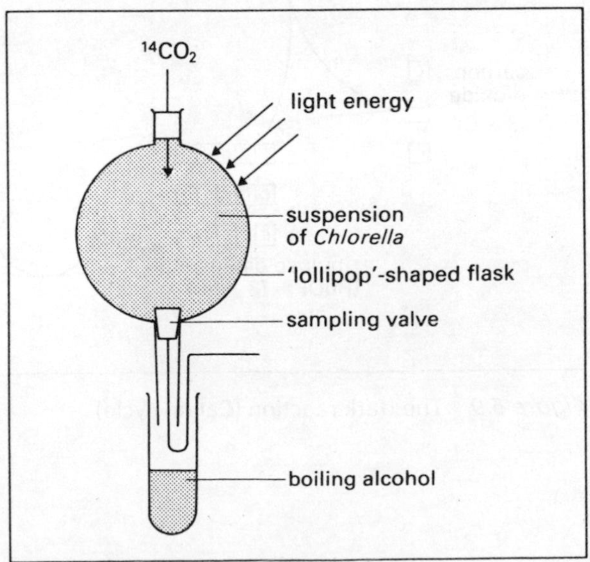

Figure 5.7 Calvin's experiment

tography (figure 5.8) to show the positions of the cell metabolites. When this chromatogram is exposed to a photographic plate the **autoradiograph** formed shows that the only metabolite to contain ^{14}C after 5 seconds is **phosphoglyceric acid (PGA)** which must therefore be the **first stable compound** formed during the dark reaction. When the same procedure is repeated at 10 seconds, 30 seconds and so on, new compounds appear labelled with ^{14}C which can also be identified. In addition, **ribulose diphosphate (RuDP)** is found to be the **carbon dioxide acceptor.** This information pieced together forms the **Calvin cycle** or dark reaction (figure 5.9) which occurs in the **stroma** of a chloroplast.

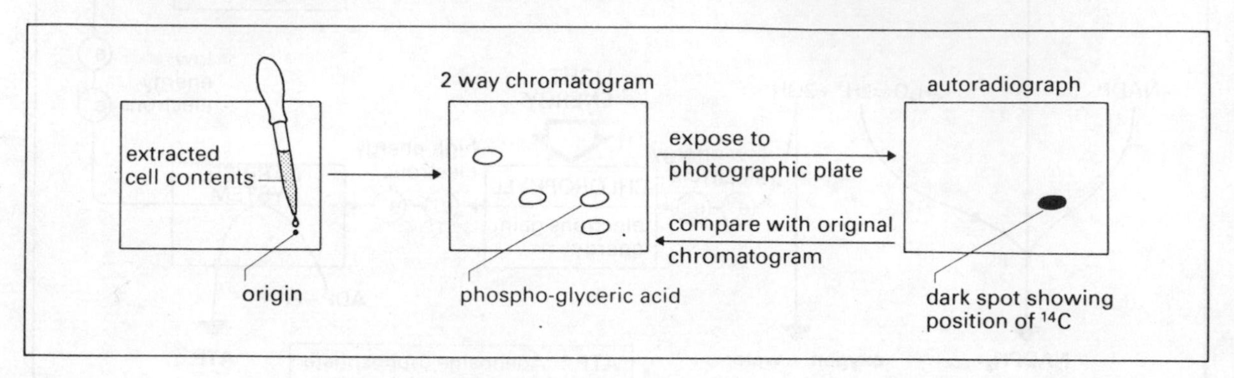

Figure 5.8 Identification of the first stable compound

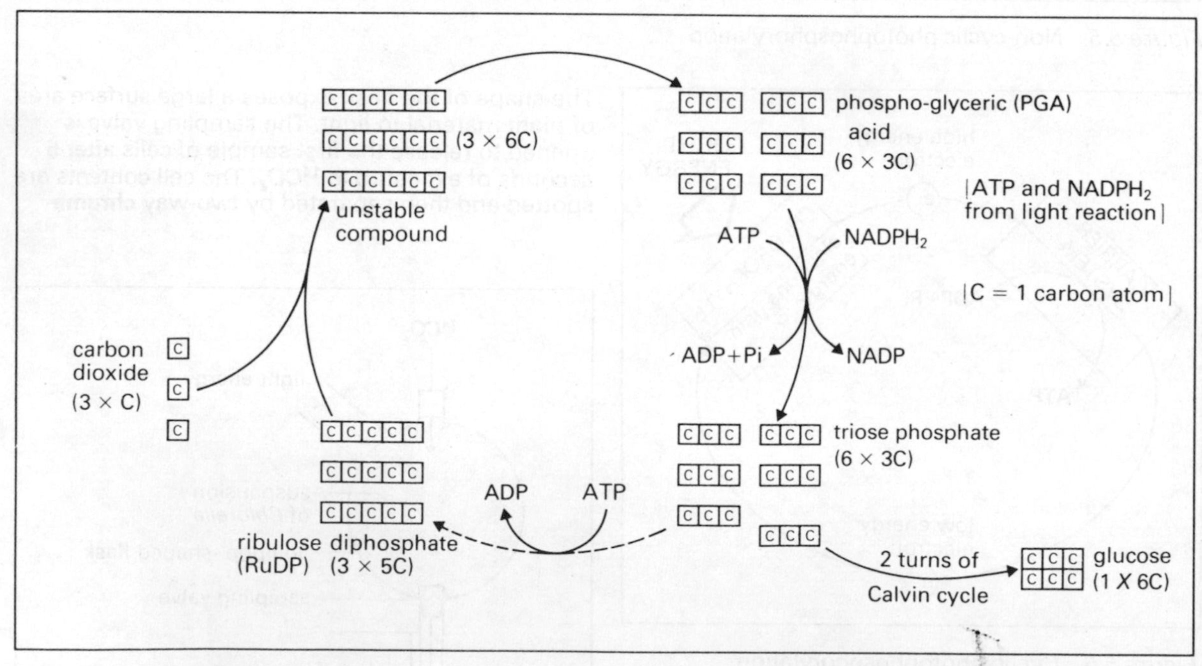

Figure 5.9 The dark reaction (Calvin cycle)

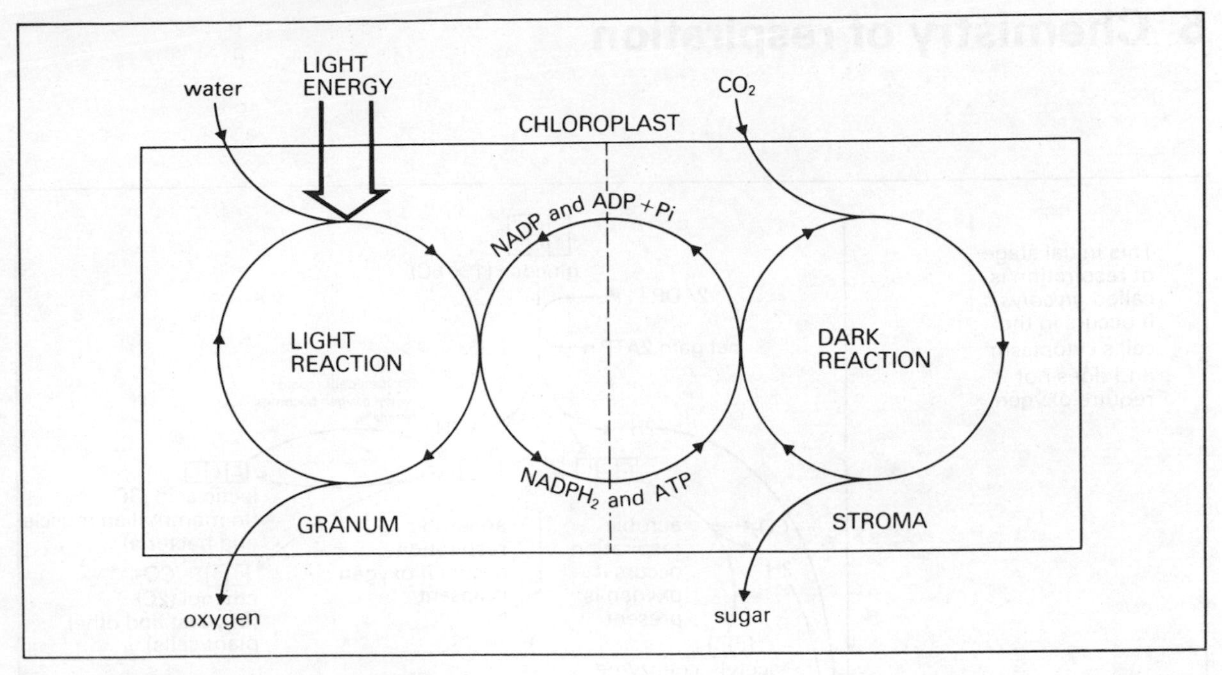

Figure 5.10 Summary of photosynthesis

Revision questions

1. What happens to a molecule of water during the light reaction?
2. Name the products of the light reaction which are essential for the dark reaction.
3. Which raw material is essential for the dark reaction and yet is not involved in the light reaction?
4. (a) Name the substances represented by the letters PGA and RuDP in the graph shown in figure 5.11.
   (b) Describe the relationship that is thought to exist between PGA and RuDP when light energy is present.
   (c) Briefly explain how the experimental results graphed here supply evidence that this relationship does exist.
5. How many times would the Calvin cycle have to turn in order to produce one molecule of sucrose ($C_{12}H_{22}O_{11}$)?

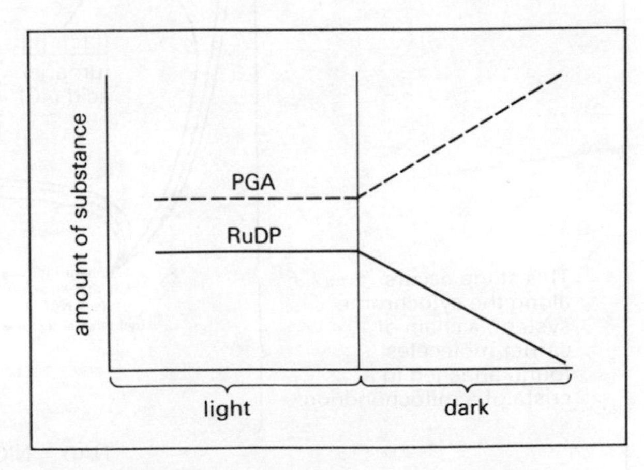

Figure 5.11

6 Chemistry of respiration

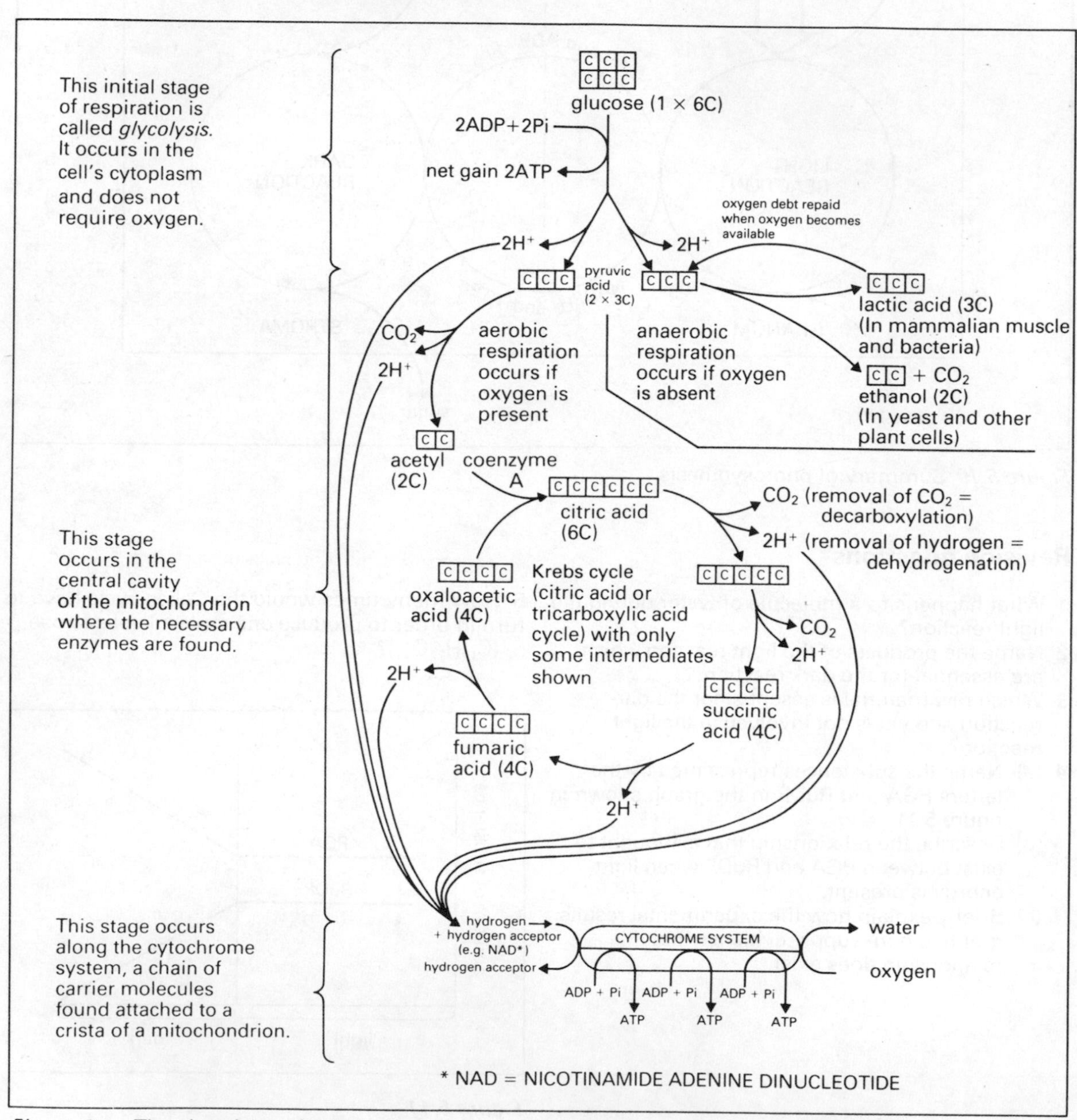

Figure 6.1 The chemistry of respiration

Oxidative phosphorylation

Each time hydrogen and electrons pass along the **cytochrome chain** (figure 6.1), the energy released forms, on average, 3 molecules of ATP. This process is called **oxidative phosphorylation** and occurs twelve times for each molecule of glucose. Thus total energy gain for one molecule of **glucose (6C)** during **aerobic** respiration = 36 ATP (formed by cytochrome system) + 2 ATP (formed during

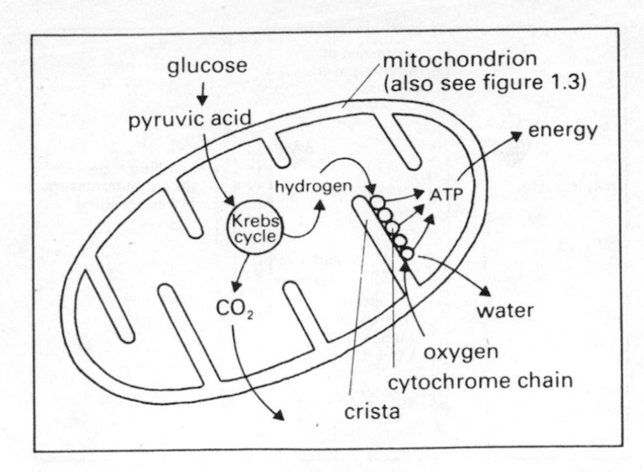

Figure 6.2 Site of aerobic respiration

glycolysis) = 38 ATP (whereas total energy gain for one molecule of glucose during anaerobic respiration = **2 ATP**).

The metabolism of other respiratory substrates (e.g. fat) enters the pathway at acetyl coenzyme A. It is important that the reactions involved in the chemistry of respiration proceed as a series of small steps catalysed by enzymes so that the energy release is gradual and under control.

Experiments to establish the sequence of intermediates.

Supplying a suspected intermediate

The discovery that **oxaloacetic acid** (when added to isolated muscle tissue) is converted to **citric acid**, enabled **Krebs** to begin working out his famous cycle (figure 6.1).

Isolating a specific enzyme

If a single enzyme can be extracted from the contents of ground up cells (e.g. liver), then the biochemist can find out which **substrate** this enzyme acts upon and which **end product** results from its action.

Using an inhibitor

Consider the biochemical pathway

$$A \xrightarrow{\text{enzyme}} B \xrightarrow{\text{enzyme X}} C \xrightarrow{\text{enzyme}} D$$

If the **inhibitor** inactivates enzyme X, then substance B accumulates and C disappears (on being converted into D). In Krebs' cycle (figure 6.1) the enzyme **succinic dehydrogenase** catalyses the conversion of **succinic acid** to **fumaric acid.** This is verified by the fact that when the inhibitor **malonic acid** is added to respiring cells, succinic acid accumulates and fumaric disappears (figure 6.3). If fumaric acid is now added to the 'blocked' system, respiration begins again. By using several inhibitors, Krebs worked out the sequence of the reactions in the cycle.

Following the radioactive tracer ^{14}C

Molecules containing atoms of radioactive carbon are introduced into the respiratory system. As a result, the first substance formed is the first to appear labelled with ^{14}C. The sequence in which the intermediate substances are formed can then

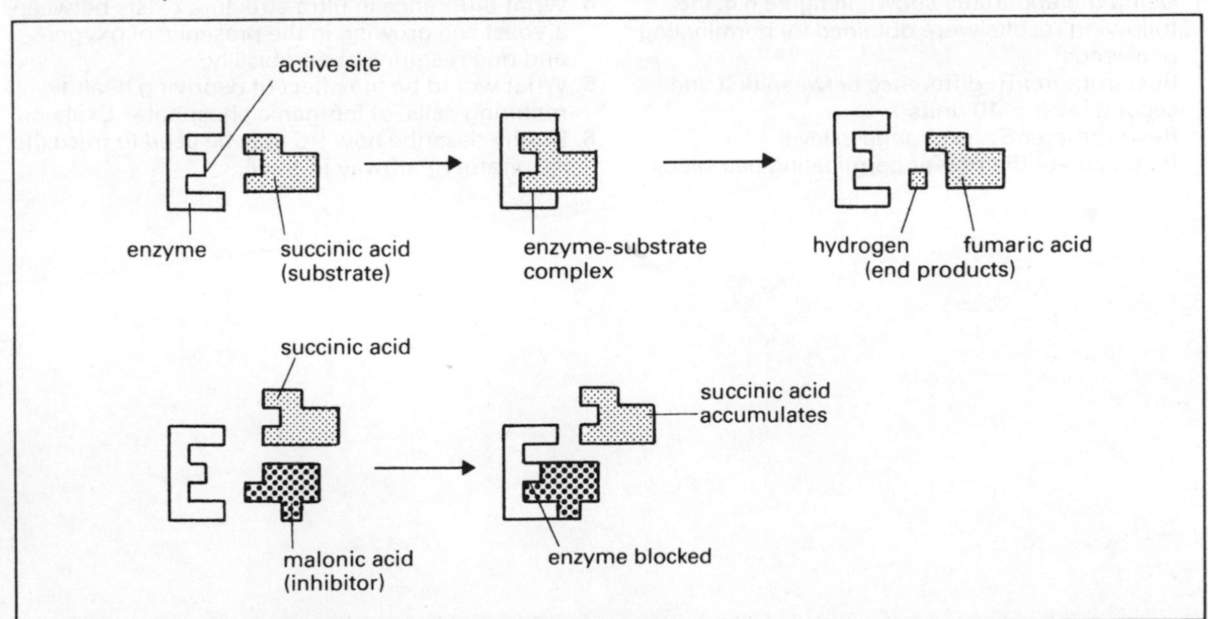

Figure 6.3 Effect of an enzyme inhibitor

be followed as ^{14}C gradually appears in turn in each compound.

Respiratory quotient

The **respiratory quotient (RQ)** for living material

$$= \frac{\text{volume } CO_2 \text{ produced}}{\text{volume } O_2 \text{ absorbed}}$$

Respiratory quotients for respiratory substrates:
RQ of carbohydrate = 1 (since vol. CO_2 produced = vol. O_2 absorbed)
RQ of fat = 0.7 } (since vol. CO_2 produced < vol.
RQ of protein = 0.6 } O_2 absorbed)

Calculation of RQ for living tissue (e.g. germinating seeds)

In R (figure 6.4), all CO_2 given out by the seeds is absorbed by the potassium hydroxide, therefore the rise of x units = the volume of O_2 absorbed by the seeds. Assuming that x units of O_2 have also been absorbed in S, then the volume of O_2 exceeds the volume of CO_2 by y units. Volume of CO_2 produced $= x - y$ units. Therefore $RQ = \frac{x - y}{x}$. Say $x = 10$ units and $y = 3$ units, then

$$RQ = \frac{10 - 3}{10} = \frac{7}{10} = 0.7$$

and it can be concluded that the respiratory substrate, being used by these seeds, is mainly fat.

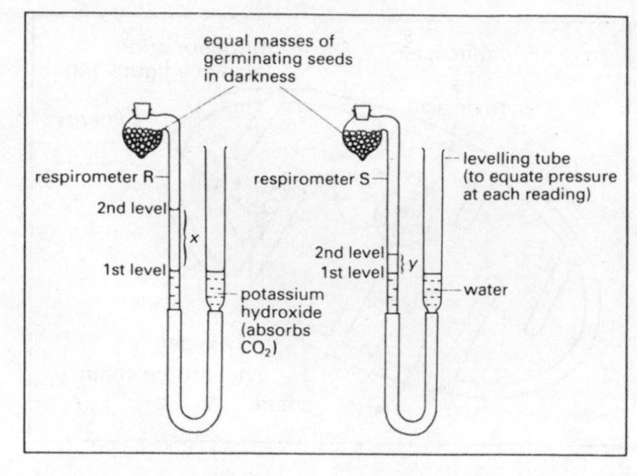

Figure 6.4 Respirometers for measuring respiratory quotient

Revision questions

1 Using the apparatus shown in figure 6.4, the following results were obtained for germinating pea seeds:
Respirometer R: difference between first and second level = 10 units
Respirometer S: no change in level
(a) Calculate the RQ for germinating pea seeds.
(b) These results indicate the use of which respiratory substrate by peas?

2 When sodium iodoacetate is added to healthy liver cells in the presence of oxygen, glucose is broken down but no pyruvic acid nor Krebs cycle metabolites are formed. (a) At which stage of the respiratory pathway does this inhibitor act? (b) Would it inhibit anaerobic respiration in yeast cells?

3 Cells poisoned by cyanide do contain pyruvic acid and Krebs cycle metabolites. (a) At which stage of the pathway does this inhibitor act? (b) Would it inhibit anaerobic respiration in yeast cells?

4 What difference in ultra-structure exists between a yeast cell growing in the presence of oxygen and one respiring anaerobically?

5 What would be the effect of depriving healthy respiring cells, of inorganic phosphate? Explain.

6 Briefly describe how ^{14}C may be used to trace the respiratory pathway in a cell.

7 Water potential

Water potential (Ψ) is the potential of a cell or solution to give out water to another cell or solution by **osmosis.** The pressure at which this water moves through a **selectively (semi-) permeable membrane** is measured in **bars.**

Osmotic potential (OP) is the potential of a cell or solution to draw water in by osmosis and exert an **osmotic pressure** also measured in bars. The OP is determined by the concentration of solute molecules present in the solution. The more solute molecules present, the higher the OP.

Water potential of pure water and aqueous solutions

Since Ψ and OP are equal and opposite, $\Psi = -$ OP. Also since the movement of water molecules is hindered by the presence of neighbouring solute molecules, pure water has the highest possible water potential. Pure water contains no solutes, therefore $OP_{\text{pure water}} = O$.

$$\text{therefore } \Psi_{\text{pure water}} = O$$

Thus 0 is the highest water potential. All solutions have lower water potentials which are **negative** values.

Water always moves from high Ψ to low Ψ. If two solutions have the same Ψ, they are **isotonic.** If they differ in Ψ then the one with the lower Ψ (high solute concentration) is **hypertonic** and the one with the higher Ψ (low solute concentration) is **hypotonic.**

Water potential of red blood cells (figure 7.1)

When the external solution is hypotonic to the cell, at the start $\Psi_{\text{solution}} > \Psi_{\text{cell}}$. Therefore water passes into the cell which swells up and bursts. This bursting and subsequent release of cell contents is called **haemolysis.** When the external solution is isotonic to the cell, at the start $\Psi_{\text{solution}} = \Psi_{\text{cell}}$. Therefore there is no net flow of water into or out of the cell. (If the OP of 0.85% salt solution = *x* bars then Ψ red blood cells = $-x$ bars). When the external solution is hypertonic to the cell, at the start $\Psi_{\text{solution}} < \Psi_{\text{cell}}$. Therefore water passes out of the cell until eventually $\Psi_{\text{solution}} = \Psi_{\text{cell}}$.

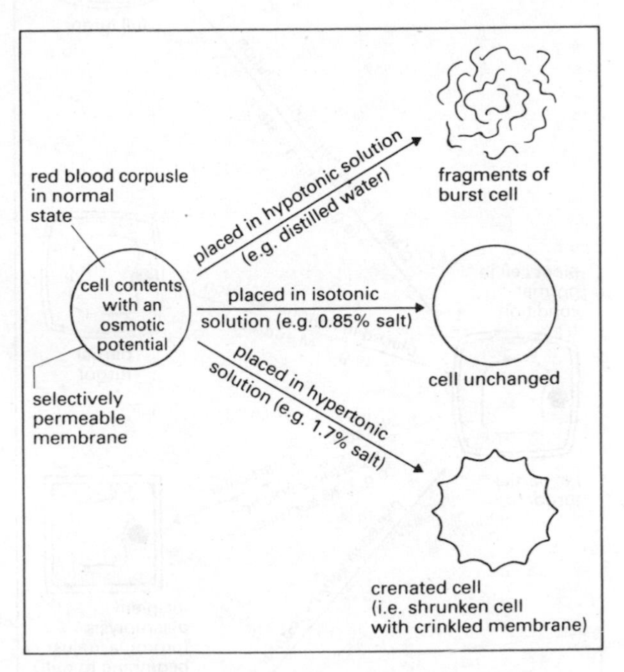

Figure 7.1 Water potential of red blood cells

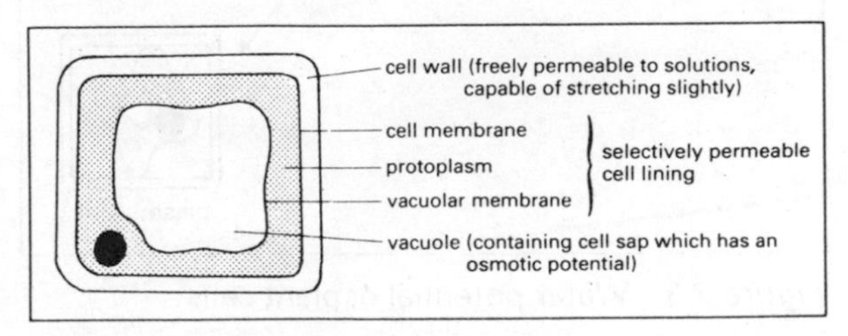

Figure 7.2 Plant cell

Water potential of plant cells

In a plant cell, the situation is complicated by the presence of a **cell wall** which, under certain conditions, presses in against the cell contents and exerts a **wall pressure (WP)**. Thus the cell's ability to give out water, i.e. water potential (Ψ_{cell}), depends on both the osmotic potential (OP_{cell}) of the vacuolar sap and the wall pressure (WP) according to the equation

$$\Psi_{\text{cell}} = WP - OP_{\text{cell}}.$$

The experiment in figure 7.3 shows the effects of various solutions on a partially turgid cell. When the external solution is hypotonic to the cell, at the start $\Psi_{\text{solution}} > \Psi_{\text{cell}}$. Therefore water passes

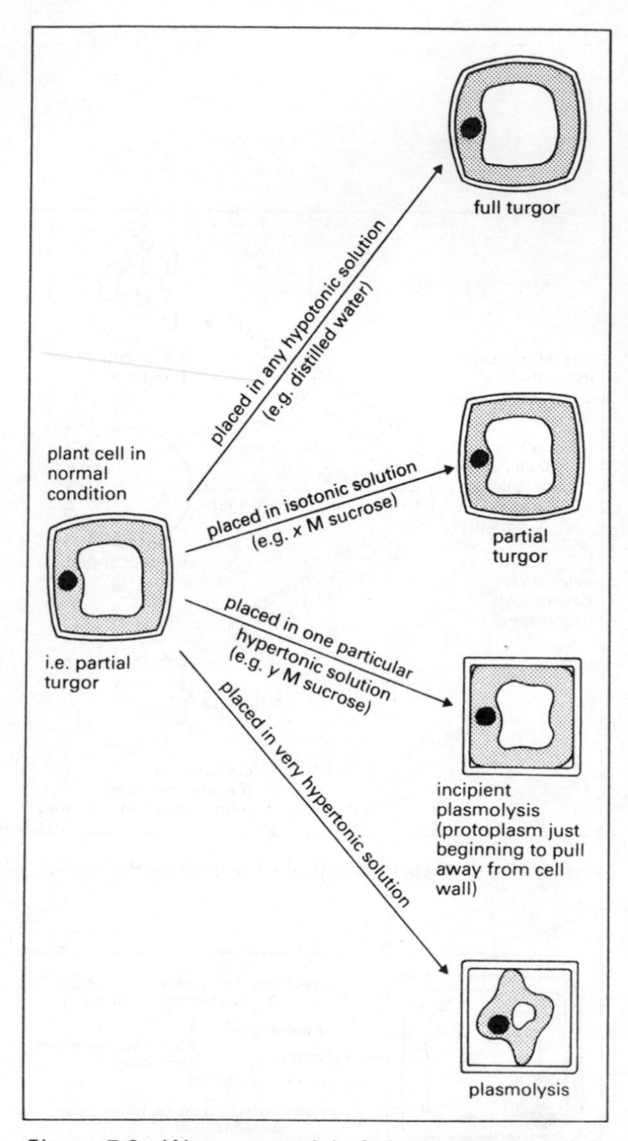

Figure 7.3 Water potential of plant cells

into the cell. OP $_{cell}$ decreases (as cell contents become more diluted). WP increases (offering an inward force against the ever expanding cell contents). Eventually WP is great enough to oppose the OP's tendency to draw more water in. Now WP = OP$_{cell}$ and since Ψ_{cell} = WP − OP $_{cell}$, Ψ_{cell} = O. The cell now has its maximum water potential. The cell is at full turgor and no more water passes in.

When the external solution is isotonic to the cell, at the start $\Psi_{solution} = \Psi_{cell}$. Therefore there is no net water flow into or out of the cell. $\Psi_{solution}$ = − OP $_{solution}$, therefore Ψ_{cell} = − OP $_{solution}$. (If *x*M sucrose has an OP = *X* bars, then Ψ_{cell} = − *X* bars.)

When the external solution is hypertonic to the cell, at the start $\Psi_{solution} < \Psi_{cell}$. Therefore water passes out of the cell. OP $_{cell}$ increases (as the cell sap becomes less dilute). WP decreases (cell contents become too small to be pressed against). Eventually WP = O, therefore Ψ_{cell} = −OP $_{cell}$. (OP $_{solution}$ = OP$_{cell}$. If *y*M sucrose has OP = *Y* bars, then OP $_{cell}$ at incipient plasmolysis = *Y* bars.)

When the external solution is very hypertonic to the cell, at the start $\Psi_{solution} \ll \Psi_{cell}$. Therefore water passes out of the cell. OP $_{cell}$ increases and WP decreases as before. Soon WP = O. The protoplast pulls away from the cell wall as water continues to pass out into the external solution. Eventually this shrinking stops when OP $_{solution}$ = OP $_{cell}$.

When placed in a hypotonic solution (high Ψ), a plasmolysed cell (low Ψ) undergoes **deplasmolysis** by taking in water. Eventually, therefore, the cell returns to a state of partial or full turgor showing that plasmolysis is a **reversible** process. Figure 7.4 summarises the relationship between OP, WP and Ψ in a plant cell.

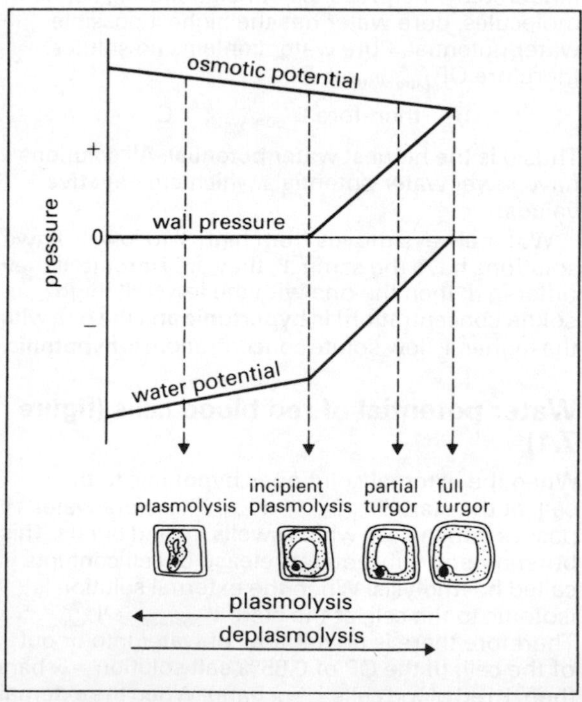

Figure 7.4 Relationship between OP, WP and Ψ in a plant cell

Revision questions

1 (a) Which solution in the osmometer experiment (see figure 7.5) has the higher OP?
(b) Which solution has the higher Ψ?
(c) In which direction will water move?
(d) What will happen to level X?

2 (a) Name structures 1 and 2 in figure 7.6 of part of a root in the ground.
(b) Which of the four lettered regions has the highest Ψ?
(c) In which direction will water move?

3 Under which of the conditions listed below would a plant cell have its (a) $\Psi = -OP$ (b) $WP = OP$ (c) $WP = 0$?

full turgor
partial turgor
incipient plasmolysis

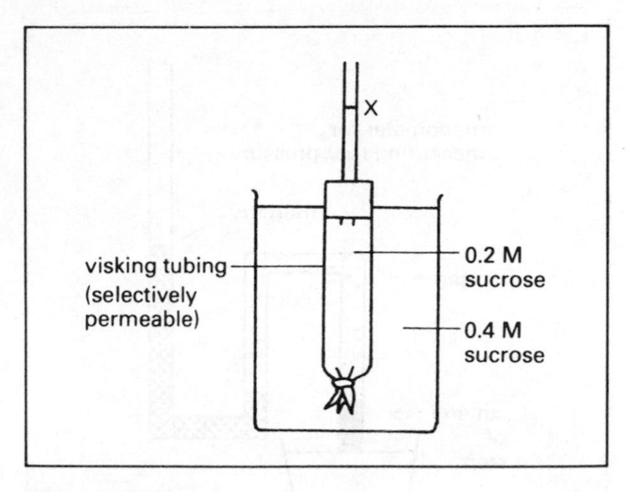

Figure 7.5

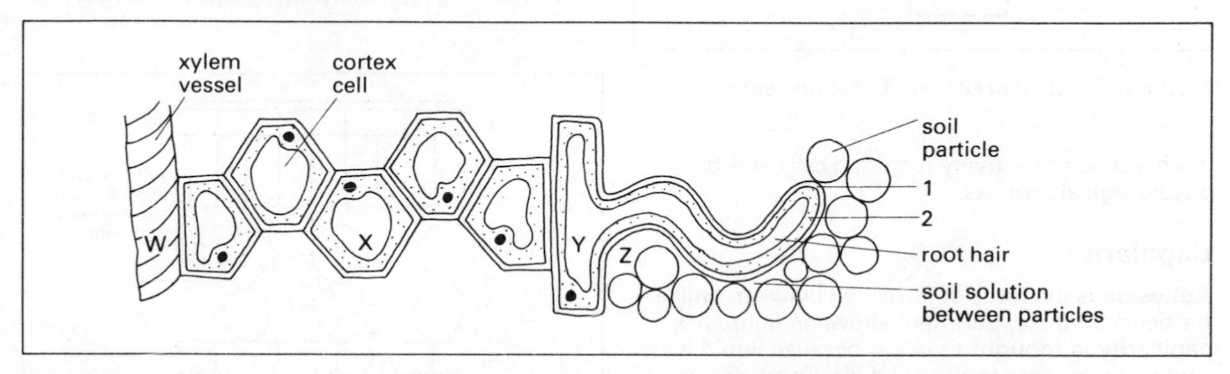

Figure 7.6

8 Water transport in woody shoots

The flow of water from a plant's roots via the stem to the leaves is called the **transpiration stream.** This ascent of water in the plant is brought about by three factors.

Root pressure

The movement of water shown in figures 8.1 and 8.2 is thought to be caused by osmosis occurring along the **water potential gradient** from the soil solution (high Ψ) via the root cells to the xylem vessels (low Ψ). Since root pressure depends upon

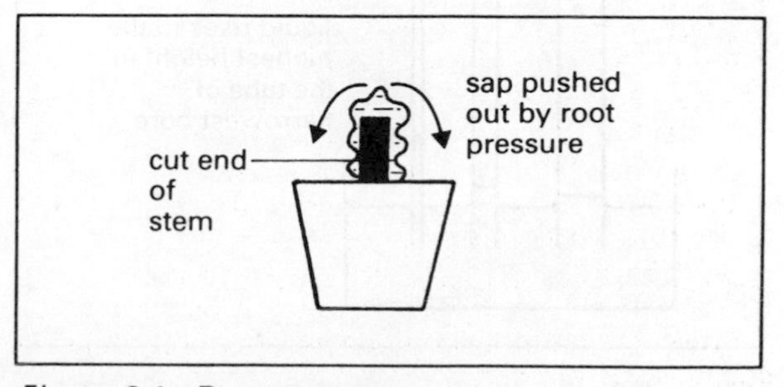

Figure 8.1 Root pressure

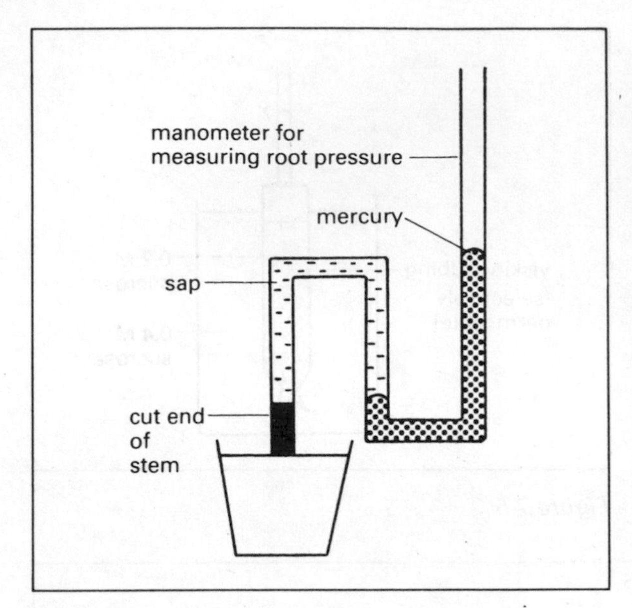

Figure 8.2 Measurement of root pressure

the presence of actively respiring cells, it is a **physiological** process.

Capillarity

Adhesion is the force of attraction between **unlike** particles. In the experiment shown in figure 8.3, capillarity is thought to occur because liquid water molecules adhere to the solid glass particles. In the narrowest tube, the largest relative surface area of glass is in contact with water, therefore most adhesion occurs. Similarly in a plant, **xylem vessels** are threadlike tubes of very fine bore and therefore water adheres to the tubes and ascends partly by capillarity. Capillarity does not depend on living cells and is a physical process.

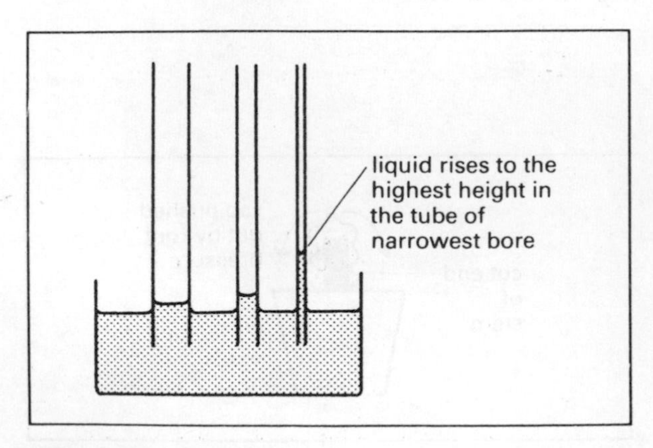

Figure 8.3 Capillarity

Transpiration pull

As a result of transpiration, water passes along a water potential gradient in a leaf (figure 8.4). As water is, in turn, drawn from the xylem vessels, a pull is exerted on the **continuous column** of water all the way down the plant.

Cohesion is the force of attraction between **like** particles. According to the Dixon-Joly cohesion theory, the great cohesive strength between the water molecules in a xylem vessel and their adhesion to the vessel wall, enables a thin tense column of water to be successfully pulled to the top of the tallest tree without snapping. The diameter of a tree decreases during periods of maximum transpiration. This demonstrates that the water in the xylem vessels is under **tension** (resulting in an inward pull because of adhesion). Transpiration pull is almost entirely a physical process with only the transpiring leaf cells playing a physiological role.

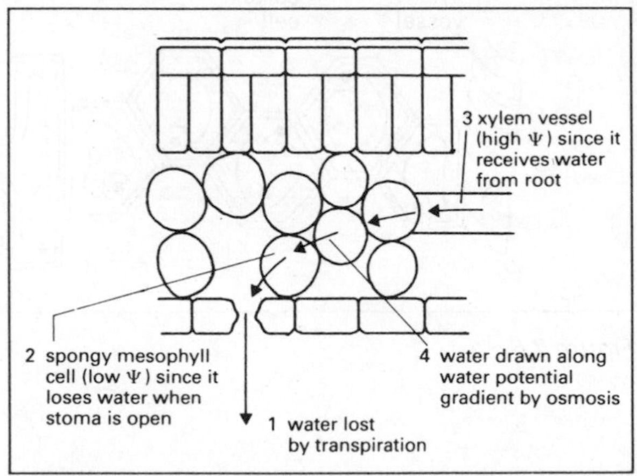

Figure 8.4 Water potential gradient in leaf

Revision questions

1 Explain why root pressure stops when a root is deprived of oxygen.

2 What is meant by the term capillarity?

3 (a) Explain why the stem of the leafy shoot shown in the experiment must be cut under water (see figure 8.5).
(b) By what process is water lost from the leafy shoot?
(c) To what important theory does this experiment lend strong support?

4 (a) Describe what happens to the tree's diameter during the 24 hour period shown in the graph in figure 8.6.
(b) Briefly explain why the process of transpiration is held responsible for this effect.

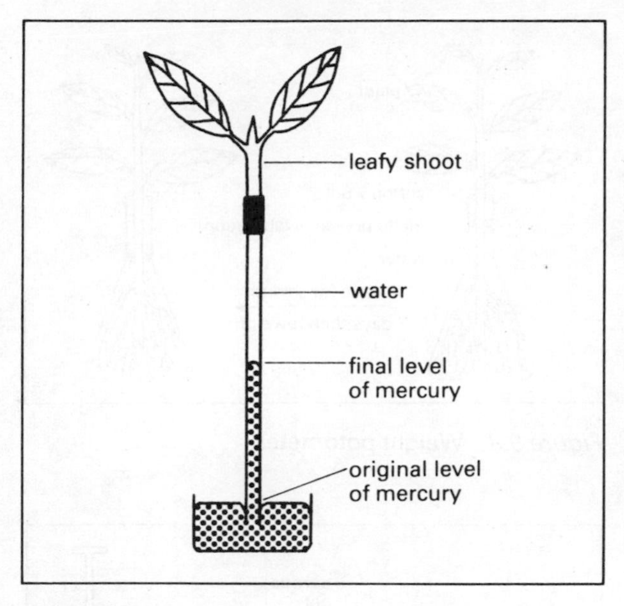

Figure 8.5

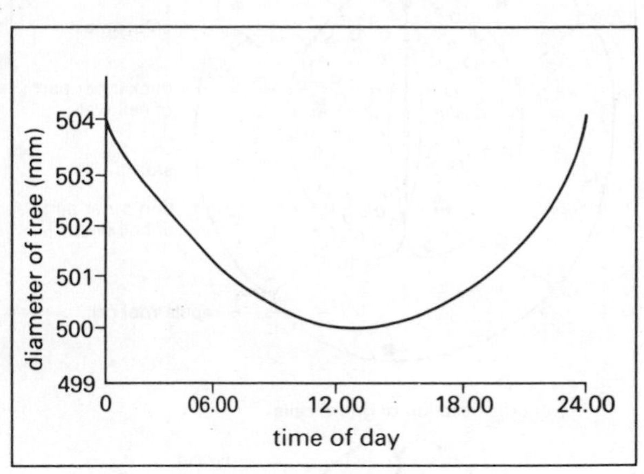

Figure 8.6

9 Transpiration

Transpiration is the process by which water is lost by evaporation from the aerial parts of a plant. Most transpiration occurs through holes in the leaves called **stomata.** When the stomata are open the plant is able to take in carbon dioxide for photosynthesis but may run the risk of losing too much water. On the other hand, a plant with closed stomata may run out of carbon dioxide. Controlled opening and closing of stomata solves this problem.

Stomatal mechanism

When water enters a flaccid guard cell (figure 9.1) **turgor** increases. Due to its larger surface area and greater elasticity, the thin outer part of the cell wall is stretched more than the thick inner part. As the outer part of each guard cell wall bulges out, the thick inner walls of the two guard cells are pulled apart thus opening the stoma.

Factors controlling stomatal mechanism

In light, photosynthesis occurs in a guard cell and its **pH** rises as carbon dioxide is used up. In darkness, photosynthesis stops and the pH drops as carbon dioxide accumulates. These changes in pH affect the action of the enzyme **starch phosphorylase** and ultimately the stomatal mechanism as shown in figure 9.2.

Potometers and atmometers

A **potometer** is an instrument used to measure transpiration from a living plant. An **atmometer** is an instrument used to measure evaporation from a non living surface.

Weight potometer

This apparatus (figure 9.3) is used to measure the total weight of water lost by transpiration from a plant.

Comparison of weight potometer and atmometer

At the end of the experiment shown in figure 9.4, when water is added from the syringe to each flask to restore the original level, it is found that in the atmometer the amount of water required is **equal**

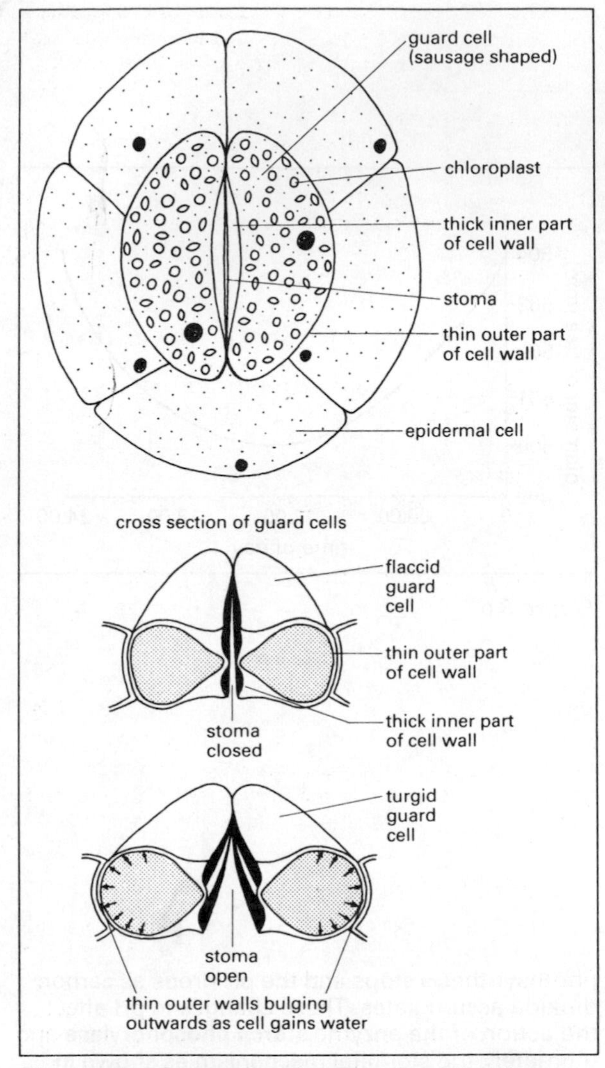

Figure 9.1 Stomata and stomatal mechanism

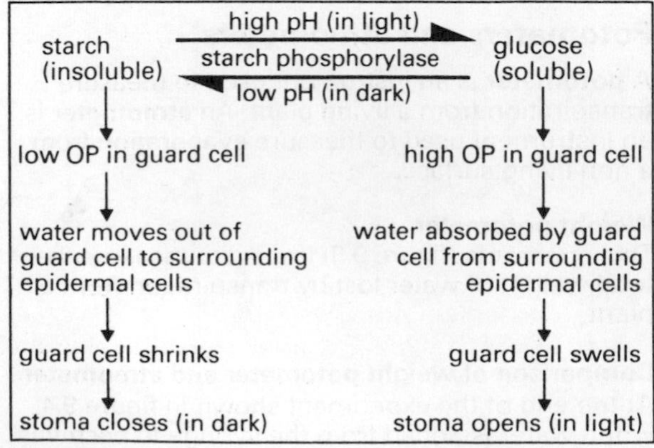

Figure 9.2 Effect of light and pH on stomatal mechanism

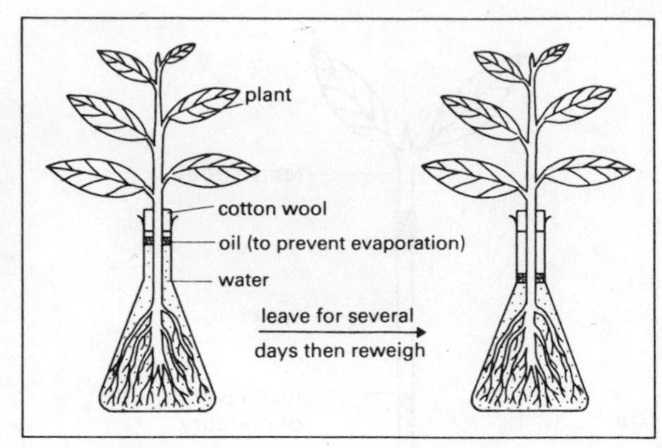

Figure 9.3 Weight potometer

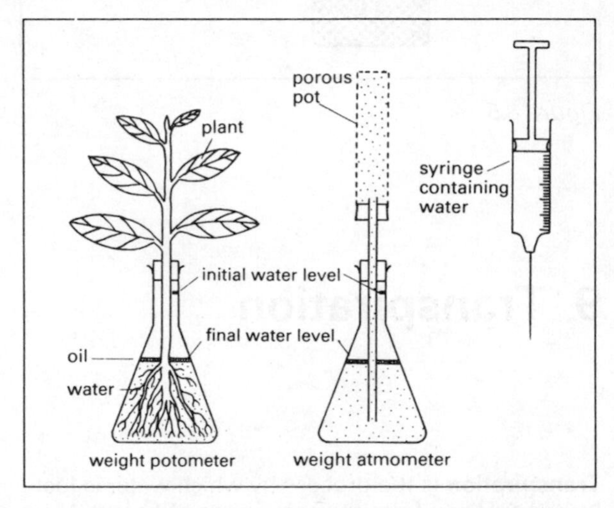

Figure 9.4 Comparison of weight potometer and atmometer

to the amount lost by evaporation whereas in the potometer the amount of water required is slightly more than that lost by transpiration. This is because the plant has used a little water for photosynthesis.

Bubble potometer

By measuring the distance travelled by the **bubble** (figure 9.5) in a given time interval, the rate of water uptake by a leafy shoot can be measured under different environmental conditions. The rate of water uptake is only approximately equal to transpiration rate since some water may be retained by the plant and used for other processes (e.g. photosynthesis). The bubble atmometer is used to measure the rate of water loss by evaporation under different conditions.

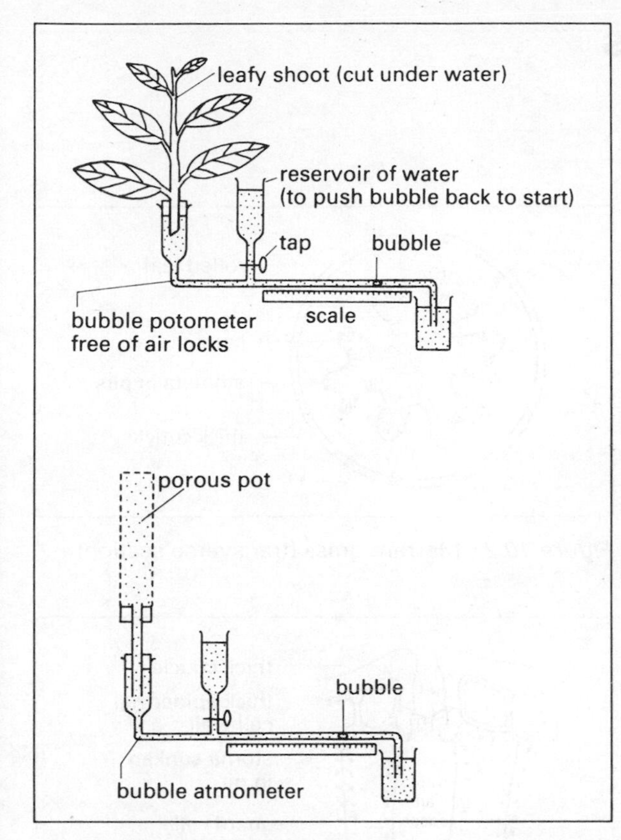

Figure 9.5 Bubble potometer and atmometer

Importance of the atmometer control
When a potometer and atmometer are subjected to the same conditions, the changes in the rate of evaporation from a plant and a purely physical system can be compared. In bright light, for example, both may show a rapid rate of water loss but in darkness only the potometer will show a drastic decrease in rate of water loss (due to stomatal closure). Thus the presence of the atmometer indicates when the potometer is acting as a free evaporator and when it is affected by physiological factors such as photosynthesis and stomatal closure.

Factors affecting transpiration rate

Temperature
Transpiration rate increases with increase in **temperature** due to faster evaporation rate of water molecules.

Humidity (amount of water vapour in air)
The air in the air spaces of a leaf is extremely humid and therefore water vapour normally diffuses out of the plant. However when the **humidity** of the external air is high, transpiration rate decreases and stops completely in saturated air which cannot accept any more water vapour.

Condition of the air
Wind increases transpiration rate because the air outside the stoma is continuously replaced with drier air which accepts water vapour from the plant. In still air transpiration decreases since the air outside the stoma becomes humid. **Air pollution** can decrease transpiration by causing blockage of stomata.

Air pressure
Transpiration rate increases as **air pressure** decreases.

Soil water
When **soil water** is in short supply, the turgidity of the guard cells cannot be maintained. The stomata therefore close and transpiration rate decreases.

Light
Stomata open in light and transpiration rate increases. The reverse occurs in darkness.

Location of plant
A plant in an isolated position shows a higher rate of transpiration than a similar plant sheltered by the dense foliage of other plants.

Revision questions

1. Why is transpiration rate not exactly equal to rate of water uptake?
2. Figure 9.6 shows a leafy shoot connected to a potometer and covered by a small metal box. (a) Name two external factors altered by the presence of the box. (b) Briefly describe the effect of each of these on transpiration rate.

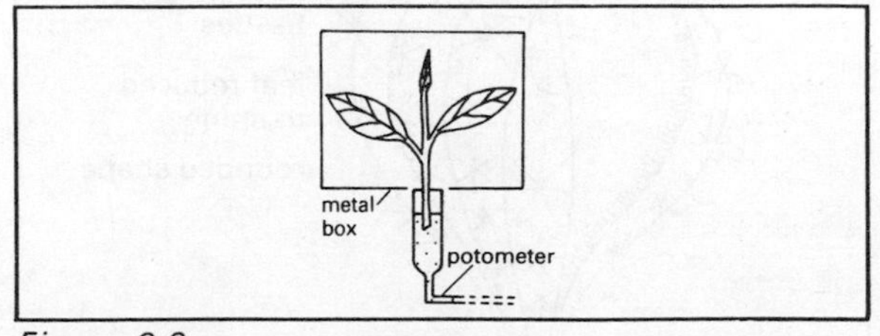

Figure 9.6

3. State two precautions which must be taken when setting up a bubble potometer.
4. Explain why a plant's stomata may close during a hot sunny afternoon only to reopen a short while later.
5. Which external factor only experienced by high altitude plants causes an increase in transpiration rate?

10 Water balance in plants

Mesophytes

These are 'normal' plants which live in habitats where water is in abundant supply and excessive transpiration does not occur. During winter, perennial mesophytes are often unable to obtain water from the frozen ground. They do not suffer, however, because they shed their leaves in autumn and therefore lose minimum water by transpiration during the winter.

Xerophytes

These plants live in habitats where a mesophyte would not survive because its transpiration rate would be excessively high. Such habitats either have hot, dry conditions and lack of soil water (e.g. desert) or exposed, windy conditions (e.g. moorland). Xerophytes achieve drought resistance by exhibiting **xeromorphic** features (figures 10.1, 10.2 and 10.3).

Clearly, minimum water is lost from a small leaf possessing a reduced number of stomata and covered by a **thick cuticle.** Transpiration is further reduced if the leaf is **rolled** or **hairy** since each of these conditions traps a layer of relatively immobile moist air between the stomata and the outer atmosphere. Stomata sunken in **pits** (which trap pockets of moist air) are similarly protected. A relatively small **surface area** liable to water loss is presented by the rounded shape of many cacti

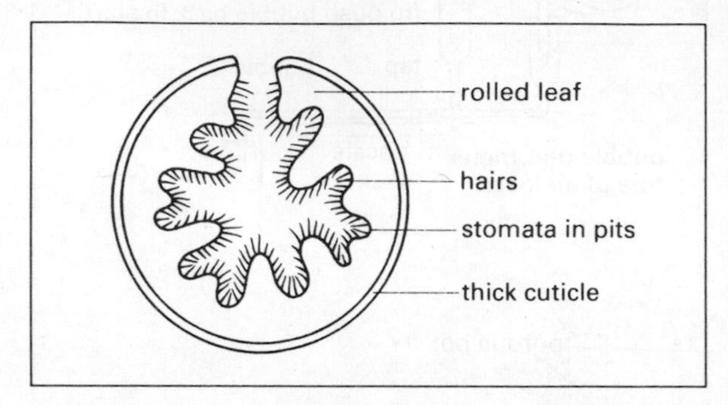

Figure 10.2 Marram grass (transverse section)

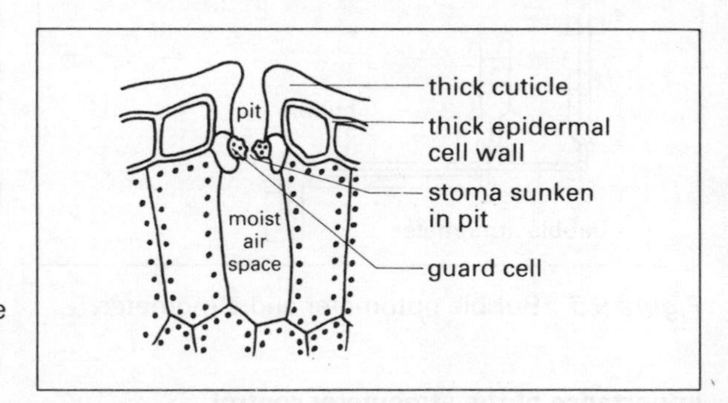

Figure 10.3 Stoma on *Hakea* leaf

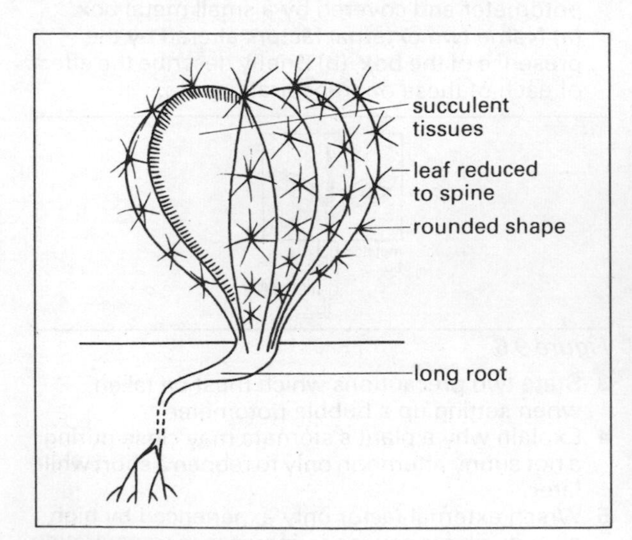

Figure 10.1 Cactus plant (part cut out)

(figure 10.1) which often also show **reversed stomatal rhythm** (i.e. stomata closed during the day and open at night). In addition they may store water in **succulent** tissues and possess **long** roots allowing absorption of subterranean water. Other desert plants possess **superficial** roots which grow parallel to the soil surface enabling them to absorb maximum water on the rare occasions that rain does fall. Some xerophytes evade drought by surviving in a highly **desiccated** state inside a hard seed coat, only germinating and going through their life cycle when water is available.

Halophytes

These are plants which live in soil often swamped with sea water. The root hairs of such plants (figure 10.4) contain sap of very high **osmotic potential** allowing water to be absorbed by osmosis from sea water. A halophyte may store water in **succulent** tissues for use when the $OP_{external} > OP_{root\ hairs}$ due to evaporation of water from mud flats at low tide.

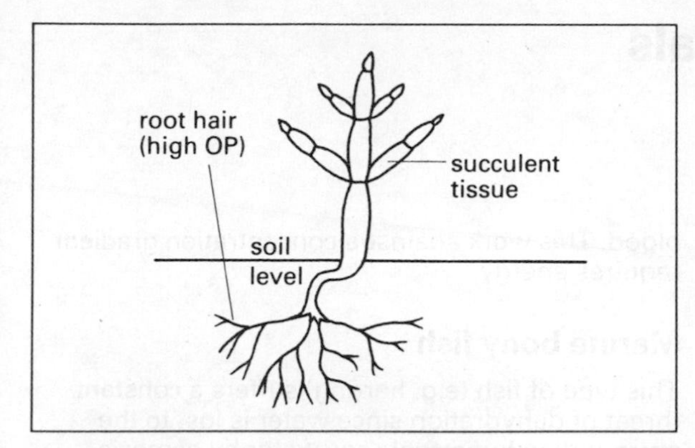

Figure 10.4 Glasswort

Hydrophytes

These are plants which live partially submerged (water-lily figure 10.5) or completely submerged (*Elodea* figure 10.6) in fresh water. A water-lily's **long petioles** allow for changes in water level. All of its stomata are found on the **upper** surface of the floating leaves thus permitting gaseous exchange with the atmosphere. Both hydrophyte types possess many **airspaces** in their tissues ensuring aeration and buoyancy. Neither possess stomata on submerged leaves. Gaseous exchange with the water occurs all over the surface of the submerged plant parts.

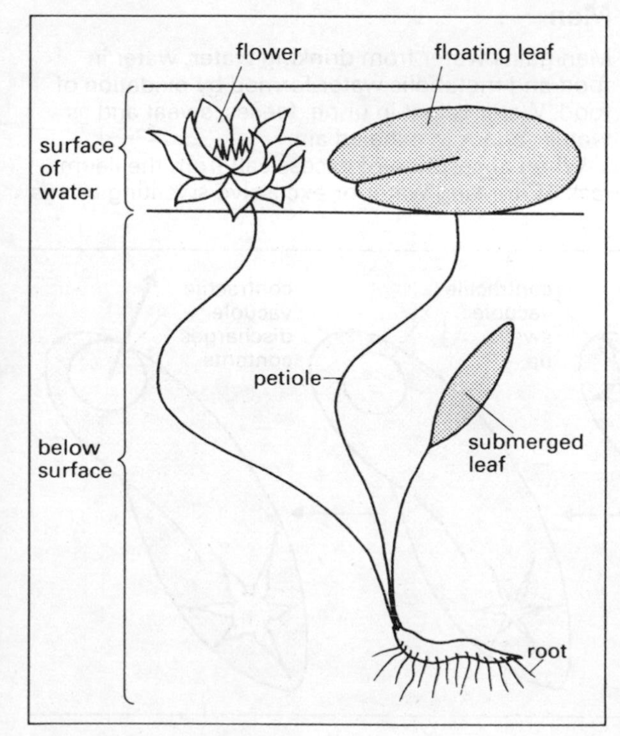

Figure 10.5 Water-lily

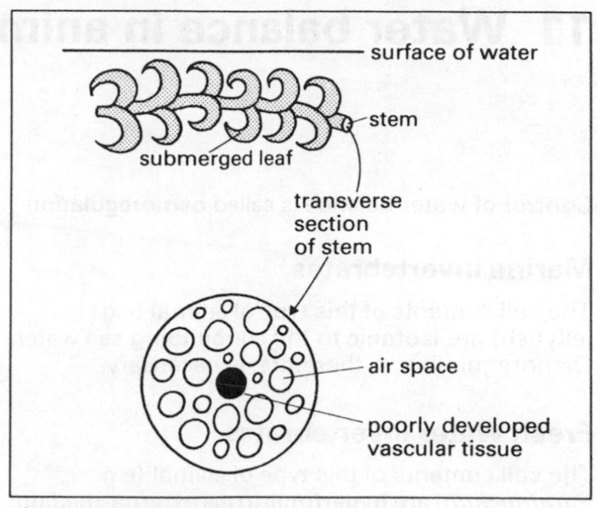

Figure 10.6 Elodea

Revision questions

1 (a) Name three xeromorphic features shown by Scots pine in figure 10.7.
(b) Briefly explain how each of these features helps the plant to resist drought.

2 What relationship exists between the water potential of a halophyte's root hairs ($\Psi_{\text{root hairs}}$) and the water potential of the soil solution (Ψ_{solution}) when (a) the soil is saturated with sea water at high tide, (b) the soil is thoroughly dried out at low tide on a warm day?

3 In which of the following plants does transpiration not occur? *Elodea,* cactus, water-lily, marram grass.

4 Why is there less vascular tissue present in the stem of a hydrophyte than the stem of a mesophyte?

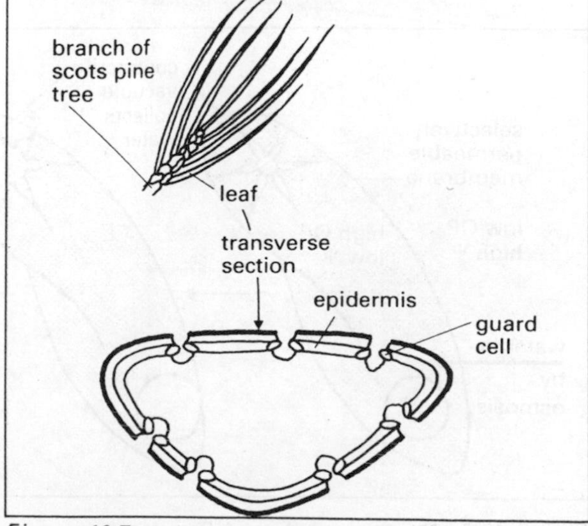

Figure 10.7

11 Water balance in animals

Control of water balance is called **osmoregulation**

Marine invertebrates

The cell contents of this type of animal (e.g. jellyfish) are **isotonic** to the surrounding sea water. Osmoregulation is, therefore, unnecessary.

Fresh water invertebrates

The cell contents of this type of animal (e.g. *Paramecium*) are **hypertonic** to the external medium and therefore water enters continuously by osmosis. To prevent the cell from swelling up and bursting, the excess water is collected in *Paramecium* by two **contractile vacuoles** which discharge their contents alternately (figure 11.1). This active process requires energy.

Fresh water bony fish

Since the body fluids of this type of fish (e.g. trout) are **hypertonic** to the external medium, water is constantly gained through the mouth and gills by osmosis (figure 11.2). This problem is overcome in two ways. The kidney contains many large **glomeruli** which allow a **high filtration** rate of blood. Much water is therefore lost as **copious urine. Chloride secretory** cells in the gills actively absorb salts from the external water (to replace those lost in urine) thus maintaining a high salt concentration in the blood. This work against a concentration gradient requires energy.

Marine bony fish

This type of fish (e.g. herring) suffers a constant threat of dehydration since water is lost to the surrounding **hypertonic** sea water by osmosis through the gills (figure 11.3). This problem is overcome in three ways. Sea water is drunk to replace losses. The kidney contains few small glomeruli. This results in a **low filtration** rate of blood and therefore little water is lost as urine. **Chloride secretory** cells in the gills, actively excrete excess salts back into the sea.

Marine cartilaginous fish

Large amounts of organic compounds (e.g. **urea**) are retained in the blood of this type of fish (e.g. shark). As a result, the total $OP_{\text{blood (salt + urea)}} = OP_{\text{sea water}}$, thus solving its water balance problem.

Man

Man gains water from drinking water, water in food and metabolic water formed by oxidation of food. Water is lost in urine, faeces, sweat and as water vapour in exhaled air.

When absorption of excess salt from the ileum, lack of drinking water or excessive sweating brings

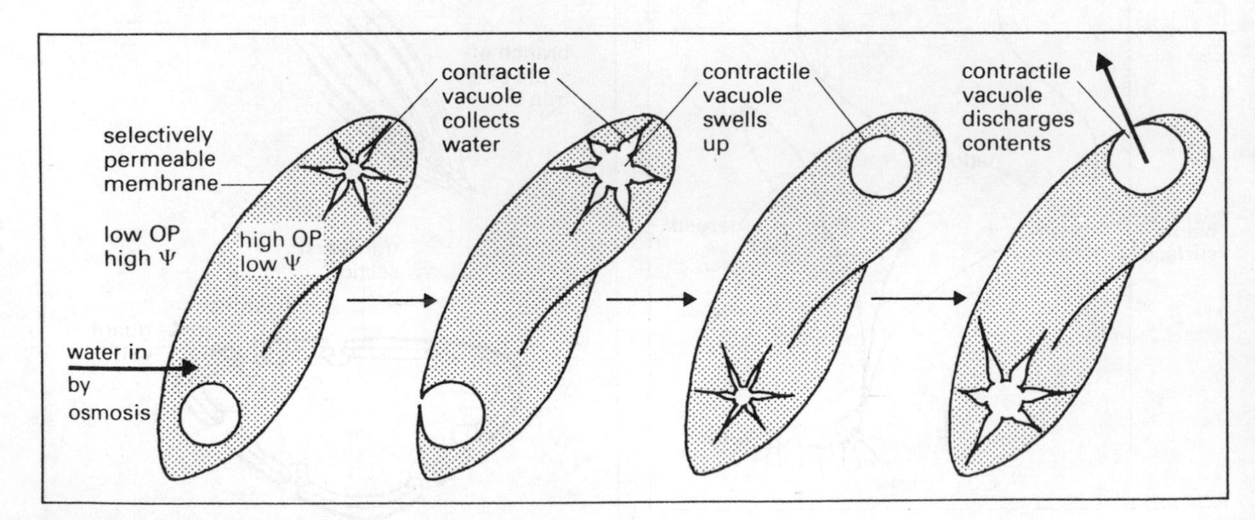

Figure 11.1 Osmoregulation in *Paramecium*

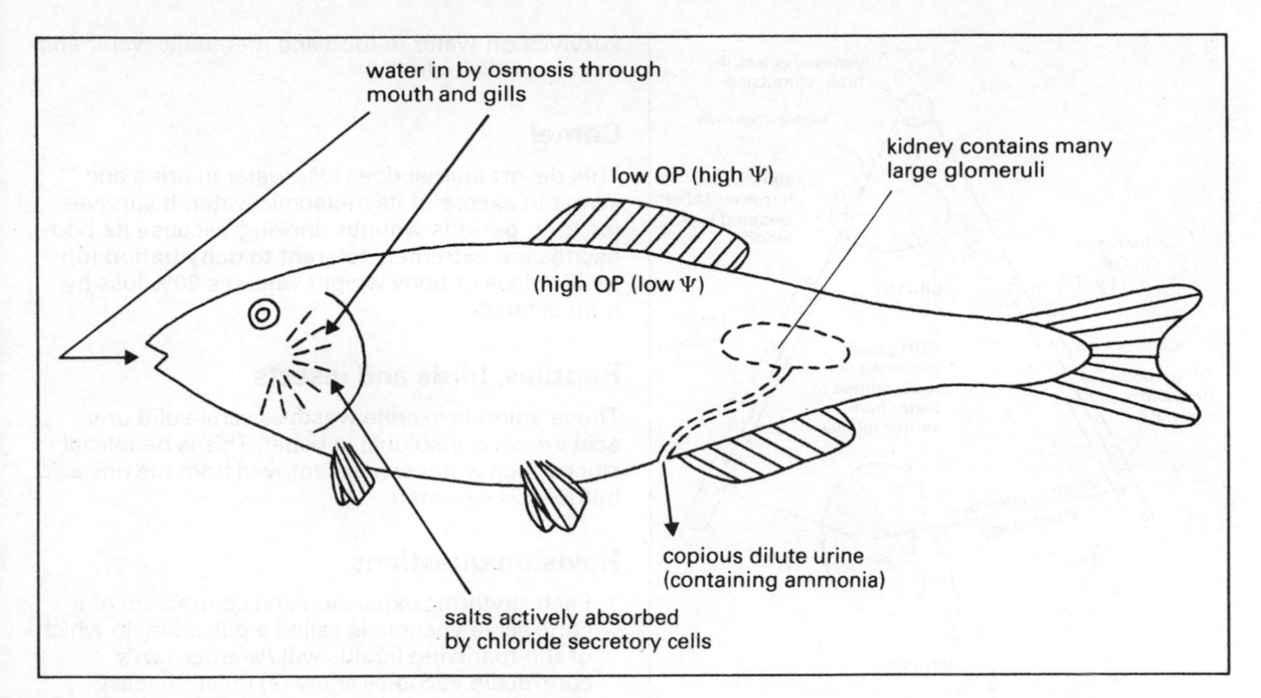

Figure 11.2 Osmoregulation in a fresh water bony fish

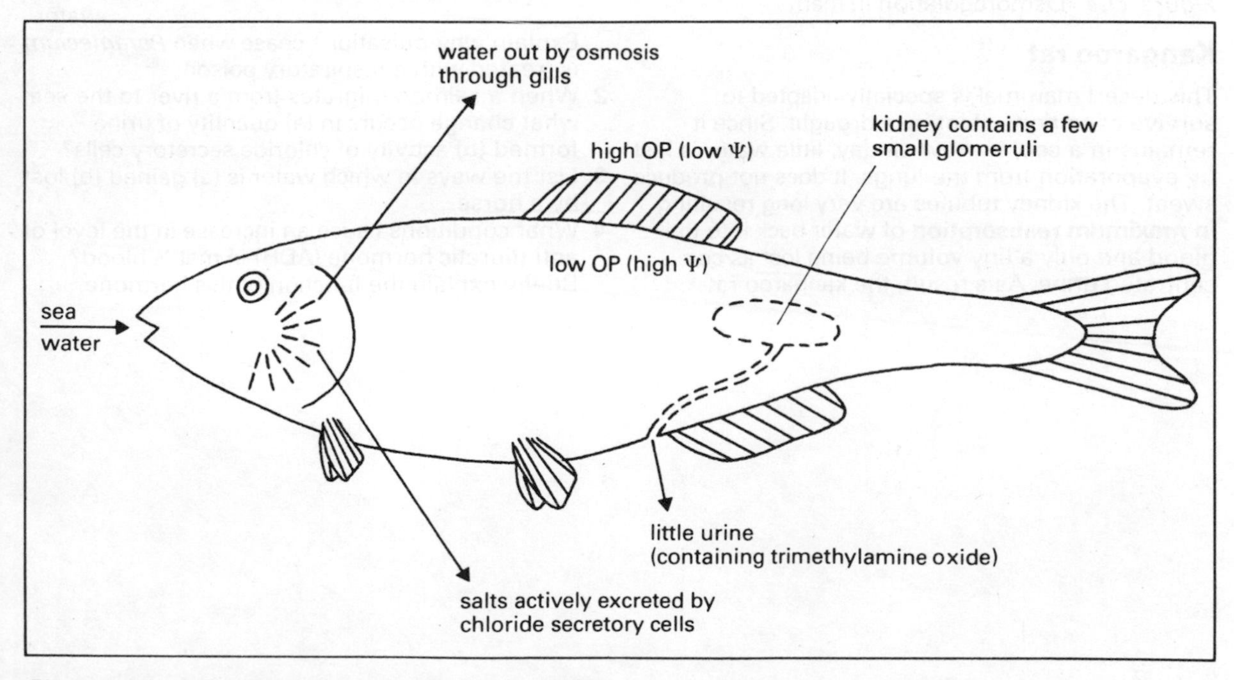

Figure 11.3 Osmoregulation in a marine bony fish

about an increase in the OP of blood, the resultant osmotic balance problem is solved by the mechanism shown in figure 11.4. Reabsorbed water lowers the OP of blood until the normal level is resumed. In addition, nerve impulses pass from the **osmoreceptors** to the cerebrum producing the sensation of thirst. Such osmoregulation is an example of **homeostasis** (see chapter 31).

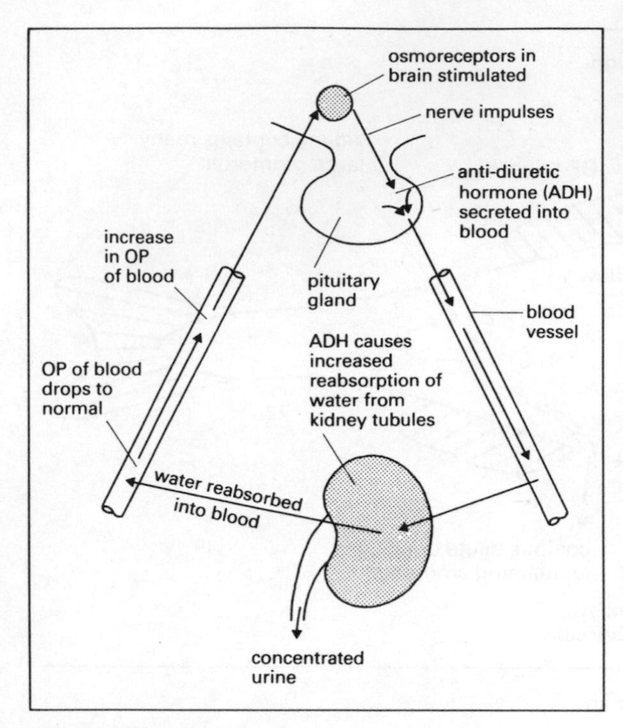

Figure 11.4 Osmoregulation in man

Kangaroo rat

This desert mammal is specially adapted to survive conditions of intense drought. Since it remains in a cool burrow all day, little water is lost by evaporation from the lungs. It does not produce sweat. The kidney tubules are very long resulting in **maximum reabsorption** of water back into the blood and only a tiny volume being lost as concentrated urine. As a result, the kangaroo rat survives on water in food and metabolic water and never needs to drink.

Camel

This desert animal does lose water in urine and sweat in excess of its metabolic water. It survives for long periods without drinking because its body tissues are extremely **tolerant to dehydration** (up to 30% loss of body weight whereas 20% loss by man is fatal).

Reptiles, birds and insects

These animals excrete waste as semi-solid **uric acid** which is insoluble in water. This is beneficial since much water can be removed from the uric acid before it is excreted.

Revision questions

1 Each rhythmic expansion and contraction of a contractile vacuole is called a pulsation. In which of the following liquids will *Paramecium's* contractile vacuoles show (a) most (b) least pulsations?
1% salt solution   0.5% salt solution   distilled water
Explain why pulsations cease when *Paramecium* is treated with a respiratory poison.

2 When a salmon migrates from a river to the sea, what change occurs in (a) quantity of urine formed (b) activity of chloride secretory cells?

3 List the ways in which water is (a) gained (b) lost by a horse.

4 What conditions cause an increase in the level of anti-diuretic hormone (ADH) in man's blood? Briefly explain the function of this hormone.

12 Translocation

Transport of the soluble carbohydrates formed by photosynthesis in a plant is called **translocation.**

From the experimental evidence presented in figures 12.1, 2 and 3 the following conclusions can be drawn. The **phloem** tissue contains the soluble carbohydrates. These carbohydrates are transported (translocated) up and down the plant in the phloem. Carbohydrates made in older leaves are translocated to growing leaves and **storage organs.**

The mechanism of translocation through the phloem is not yet fully understood but certainly seems to depend upon the **metabolic activity** of the phloem sieve tubes because any condition that affects their metabolism also affects translocation.

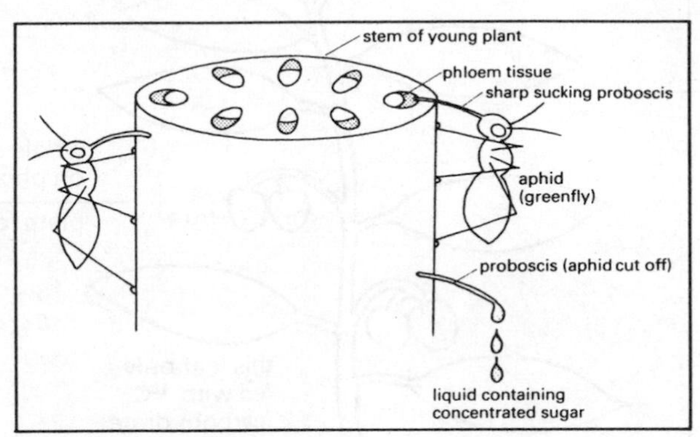

Figure 12.1 Aphids

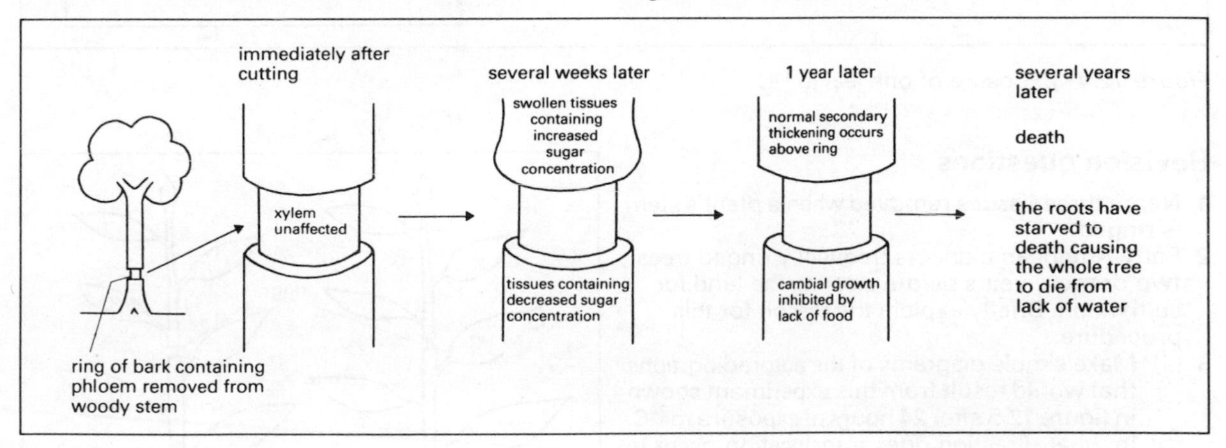

Figure 12.2 Ringing (girdling)

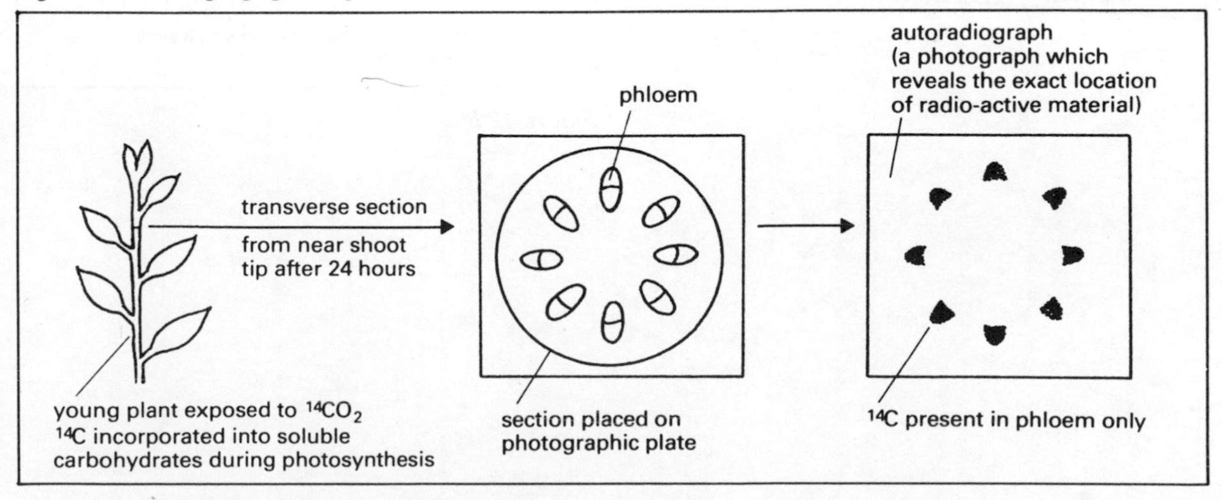

Figure 12.3 Exposure of whole plant shoot to ^{14}C

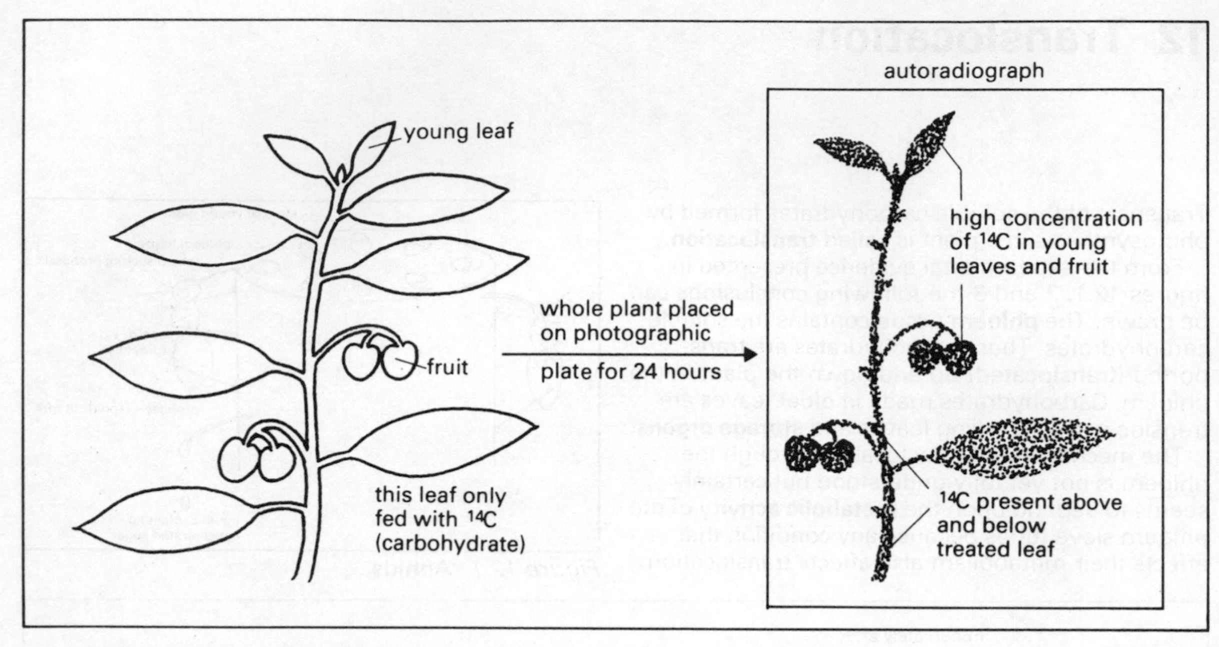

Figure 12.4 Exposure of one leaf to ^{14}C

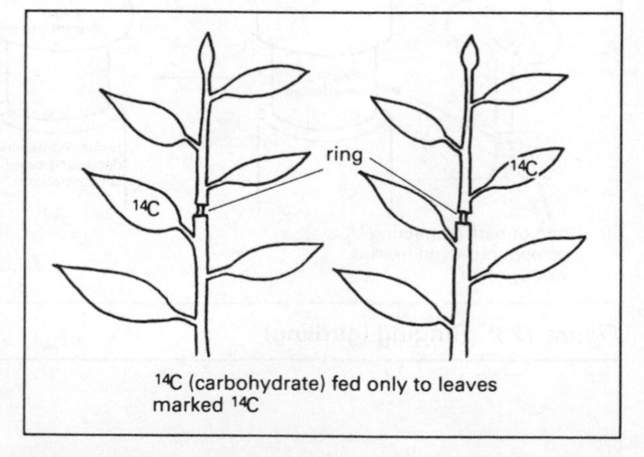

Figure 12.5

Revision questions

1 Name three tissues removed when a plant's stem is ringed.

2 Early American pioneers frequently ringed trees two or more years before clearing the land for cultivation. Briefly explain the reason for this procedure.

3 (a) Make simple diagrams of the autoradiographs that would result from this experiment shown in figure 12.5 after 24 hours of exposure to ^{14}C.

(b) In what direction does translocation occur in a plant?

13 Inorganic solutes

Van Helmont's experiment

In 1648 when Van Helmont reported the results of his experiment (figure 13.1) he did not appreciate the roles of CO_2, light energy or chlorophyll in photosynthesis but concluded instead that 77.5 kg of willow plant had arisen from water alone. Neither did he place any importance on the loss in weight of 0.05 kg by the original 100 kg of soil. We now know that this loss in weight amounts to the quantity of **nitrogen** (as nitrate) and **mineral elements** absorbed from the soil by the plant.

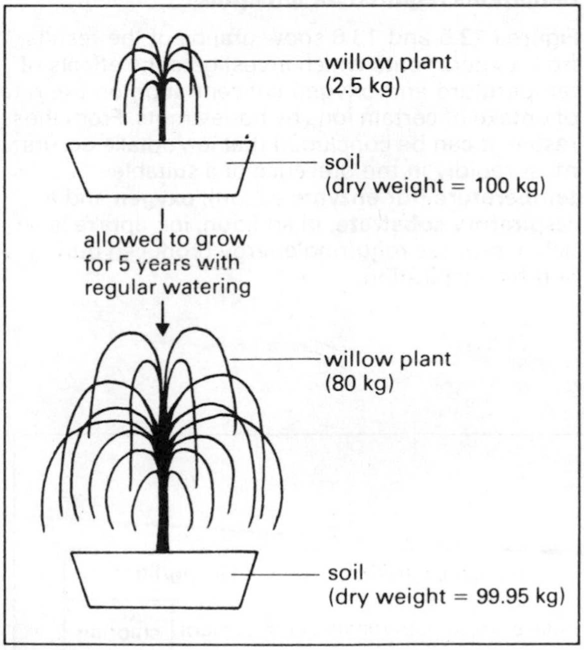

willow plant (2.5 kg)
soil (dry weight = 100 kg)
allowed to grow for 5 years with regular watering
willow plant (80 kg)
soil (dry weight = 99.95 kg)

Figure 13.1 Van Helmont's experiment

Determination of the ash content of a plant

The **ash content** of a corn plant can be determined as shown in figure 13.2. From the results it can be concluded that in addition to organic matter, a plant contains small amounts of many different inorganic **mineral elements.**

Water culture experiments

The importance of individual elements to a plant is demonstrated by **water culture** experiments

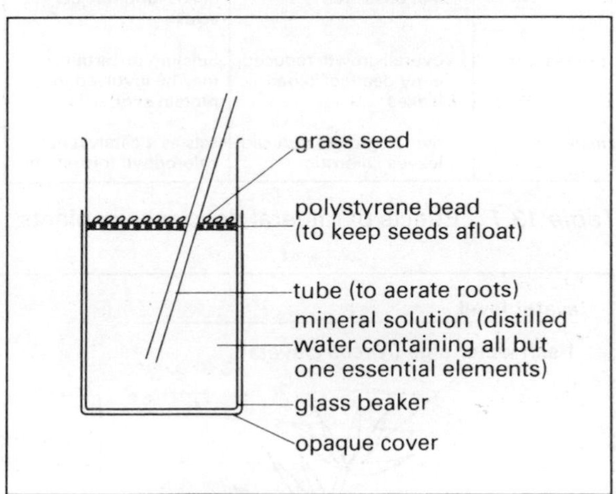

Figure 13.3 Water culture experiment

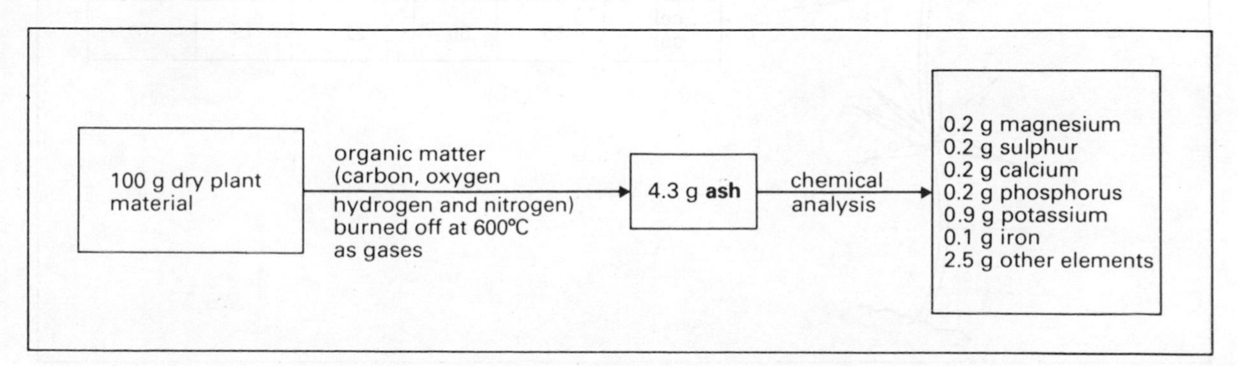

Figure 13.2 Determination of ash content of corn plant

(figure 13.3). The glass beaker is initially rinsed with concentrated nitric acid to remove traces of mineral elements and then surrounded by an opaque cover to keep out light and prevent algal growth. By omitting one element at a time, the importance of each element to the plant can be assessed (table 13.1).

Element omitted	Symptoms of deficiency	Reason for deficiency symptom (role of element)
nitrogen	overall growth reduced, leaves chlorotic (pale green or yellow), leaf bases red, roots long and thin	required for formation of protein and nucleic acids
magnesium	overall growth reduced, leaves chlorotic	required for chlorophyll formation
sulphur	overall growth reduced	required for protein formation
calcium	overall growth drastically reduced, few leaves	required for formation of original cell wall
phosphorus	overall growth reduced, leaf bases red	required for formation of ATP and nucleic acids
potassium	overall growth reduced, early death of older leaves	function uncertain, may be involved in protein synthesis
iron	overall growth reduced, leaves chlorotic	acts as a catalyst in chlorophyll formation

Table 13.1 Effects of mineral deficiency in plants

Trace elements
Elements such as **zinc, copper** and **manganese** are also required by plants in tiny amounts (probably because they act as co-factors in enzyme reactions).

Ion uptake

From the experiment shown in figure 13.4 it can be concluded that fresh water plants are able to select and accumulate certain elements (in the form of ions) within their cells to concentrations greatly in excess of the concentrations in the external environment. Such ion uptake from low to high concentration (against a concentration gradient) cannot be brought about by diffusion.

Conditions required for ion uptake

Figures 13.5 and 13.6 show graphs of the results from experiments which investigate the effects of temperature and oxygen concentration on the rate of uptake of certain ions by barley roots. From these results it can be concluded that ion uptake occurs most rapidly in the presence of a suitable **temperature** (for enzyme action), **oxygen** and a **respiratory substrate**. In addition, ion uptake is an active process requiring energy produced during aerobic respiration.

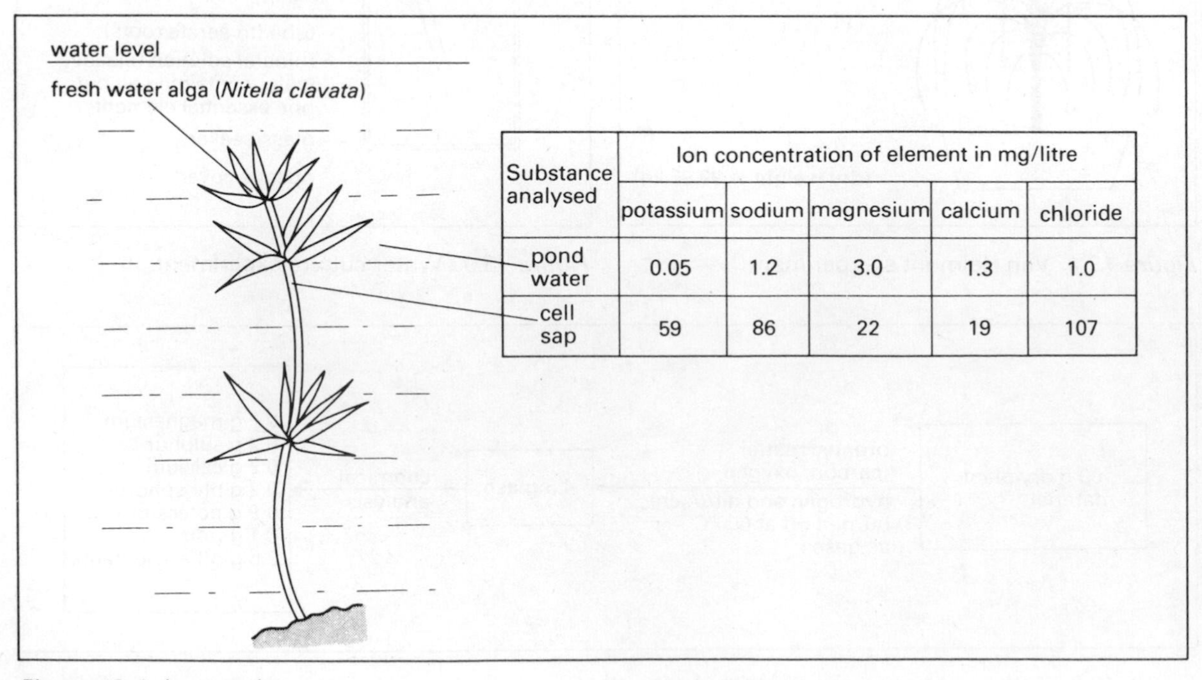

Substance analysed	Ion concentration of element in mg/litre				
	potassium	sodium	magnesium	calcium	chloride
pond water	0.05	1.2	3.0	1.3	1.0
cell sap	59	86	22	19	107

Figure 13.4 Ion uptake

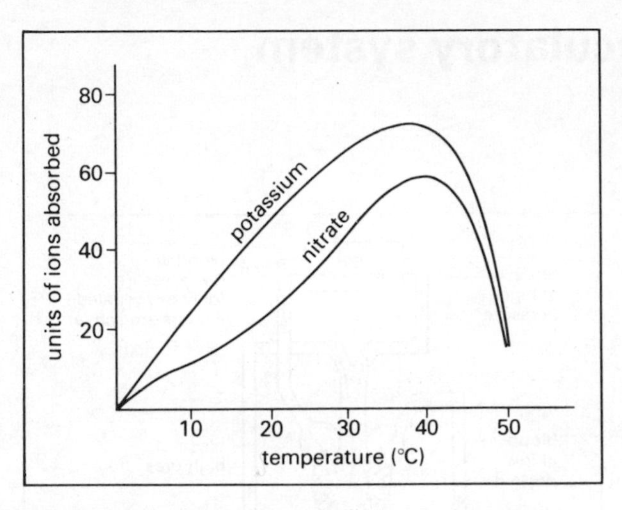

Figure 13.5 Effect of temperature on ion uptake

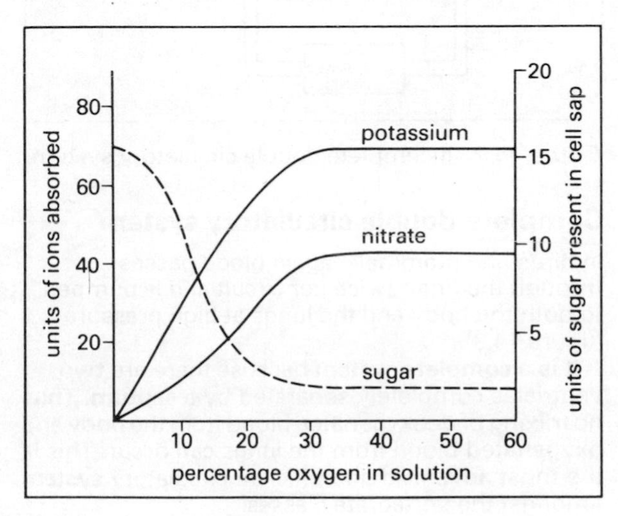

Figure 13.6 Effect of oxygen concentration on ion uptake

Revision questions

1 Describe the appearance of a chlorotic leaf.

2 (a) Name an element which, if in deficient supply to a plant, brings about chlorosis.
(b) Relate this element's role in the plant's metabolsim, to chlorosis.
(c) Chlorosis rarely becomes apparent until a few weeks after germination. Suggest the reason why.
(d) What control should be set up in a water culture experiment?

3 (a) Describe briefly how the plant shown in figure 13.7 deals with each type of ion.
(b) What conditions must be present for such cell activities to work?

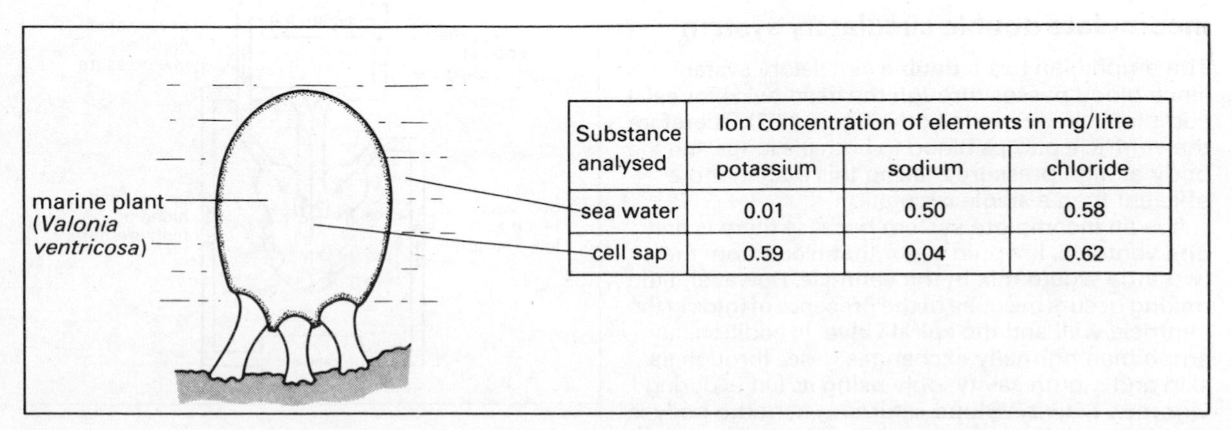

Substance analysed	Ion concentration of elements in mg/litre		
	potassium	sodium	chloride
sea water	0.01	0.50	0.58
cell sap	0.59	0.04	0.62

Figure 13.7

14 Types of vertebrate circulatory system

Single circulatory system

The fish has a **single** circulatory system since blood only passes through the heart once for each complete circuit around the body (figure 14.1). In any circulatory system a drop in pressure occurs when blood passes through a **capillary bed** (a network of narrow tubes which offer a resistance). Thus in a fish, blood flows to the gills at high pressure but then on to the body at low pressure. This is a primitive and relatively inefficient method of circulation compared to other vertebrate classes.

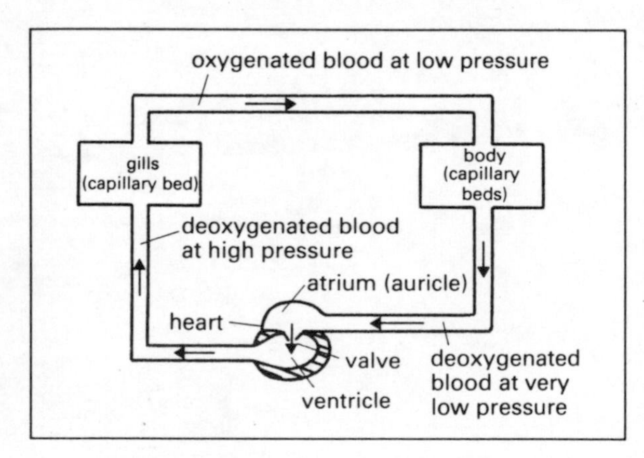

Figure 14.1 Single circulatory system

Incomplete double circulatory system

The amphibian has a **double** circulatory system since blood passes through the heart twice for each complete circuit of the body (figure 14.2). Therefore the ventricle pumps blood to both the lungs and the body at high pressure making this system more efficient than a single circulation.

It is an **incomplete** system because there is only **one** ventricle. It would seem that blood from the two atria would mix in the ventricle. However, little mixing occurs because of the presence of **folds** in the ventricle wall and the **spiral valve.** In addition, an amphibian normally exchanges gases through its skin and mouth cavity, only using its lungs during vigorous activity. Blood returning from the body is usually **partially oxygenated.**

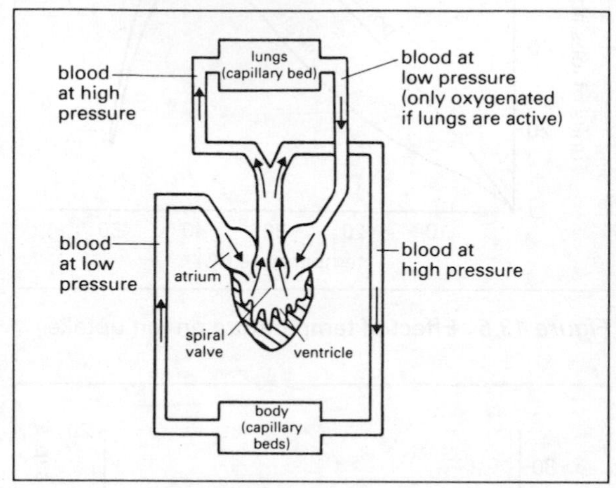

Figure 14.2 Incomplete double circulatory system

Complete double circulatory system

In birds and mammals, again blood passes through the heart twice per circuit and is pumped to both the body and the lungs at high pressure (figure 14.3).

It is a **complete** system because there are **two** ventricles completely separated by a **septum.** Thus no mixing of deoxygenated blood from the body and oxygenated blood from the lungs can occur. This is the most advanced and efficient circulatory system amongst the vertebrate classes.

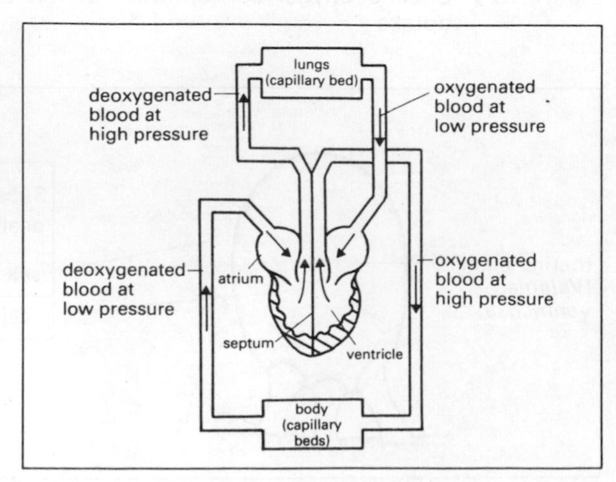

Figure 14.3 Complete double circulatory system

Revision questions

1 Copy and complete the blanks in the table.

Animal class	Type of circulation	Number of chambers in heart	Pressure of blood arriving at skeletal muscles	Evolutionary level of system
fish				
	incomplete double			intermediate
mammal				

Table 14.1

2 When an amphibian gains oxygen by cutaneous respiration, which atrium receives the oxygenated blood?

3 How many times must a red blood cell in the pulmonary vein of a mammal pass through the heart before reaching the lungs?

4 (a) Which type of circulatory system is shown in figure 14.4?
(b) Name an organism which possesses such a system.
(c) Name blood vessels A and B.
(d) Will blood pressure at point C be high or low?

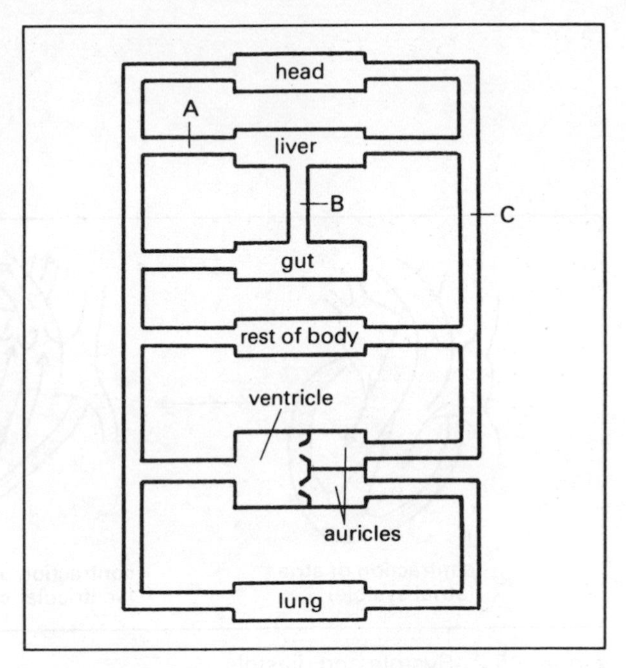

Figure 14.4

15 Control of vertebrate circulation

The circulatory system consists of a muscular pump, the heart, and a series of blood vessels, arteries, capillaries and veins, through which the heart pumps blood to all parts of the body.

Origin of heartbeat

Approximately seventy times per minute, a wave of excitation spreads from the **pacemaker** (figure 15.1) over the atria causing them to contract. This also stimulates the **A-V node** which then sends electrical signals at great speed along **conducting fibres** situated in the ventricle walls. This causes the ventricle muscles to contract simultaneously and pump blood into the arteries at high pressure. The entire contraction operation is **systole** (figure 15.2) whereas the period of relaxation between heart-beats is **diastole.** This rhythmic beating originates in the heart (a frog's heart removed from the body continues to beat for some time). However the rate of beating is controlled by nerves and hormones.

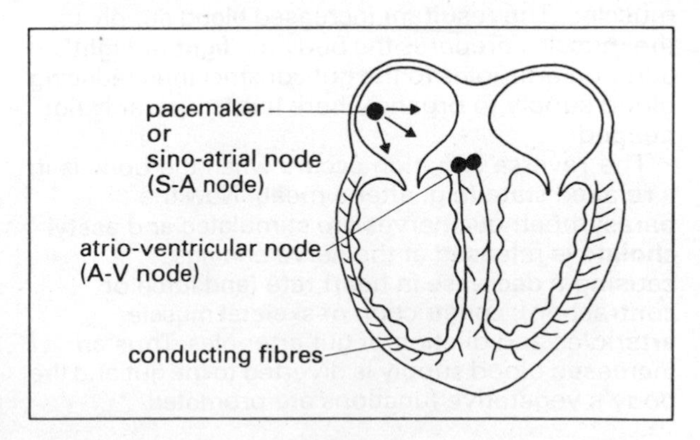

Figure 15.1 Origin of heartbeat

Nervous control of circulation

Nervous control is carried out by the **autonomic nervous system** which consists of the **sympathetic**

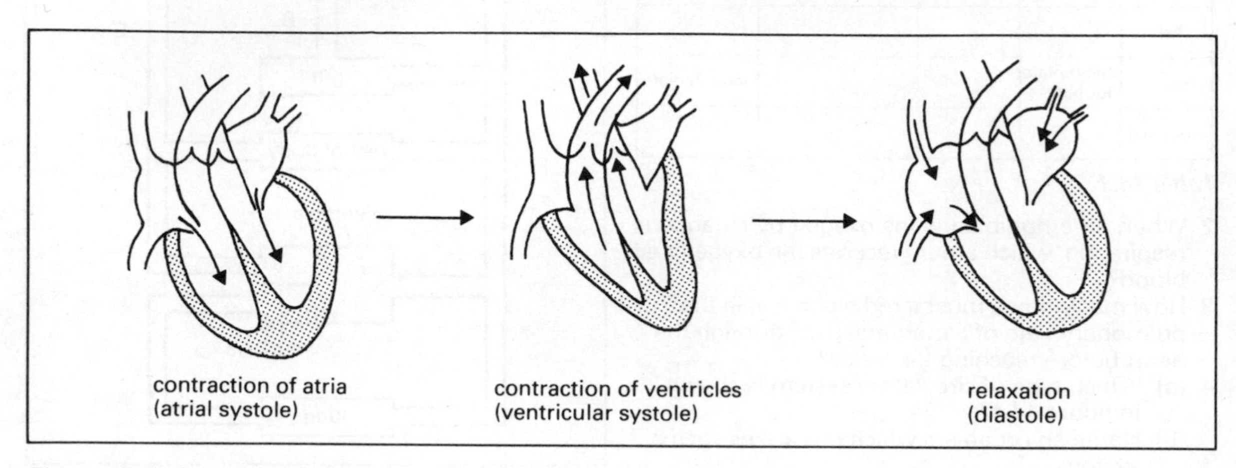

Figure 15.2 Systole and diastole

and **parasympathetic** nervous systems (figure 15.3). These act below the level of consciousness.

During stress or danger the sympathetic nerves are stimulated and the chemical transmitter, **noradrenalin,** is released at the nerve endings. This causes increase in heart rate (and force of contraction) and dilation of arterioles to skeletal muscles. The resultant increased blood supply to the muscles prepares the body for **'fight or flight'**. In addition arterioles to the gut constrict thus reducing blood supply to organs where it is temporarily not needed.

The reverse situation occurs when the body is in a relaxed state (e.g. after a meal). Now the parasympathetic nerves are stimulated and **acetyl choline** is released at the nerve endings causing a decrease in heart rate (and force of contraction), constriction of skeletal muscle arterioles and dilation of gut arterioles. Thus an increased blood supply is diverted to the gut and the body's vegetative functions are promoted.

Hormonal control of circulation

Stimulation of the sympathetic nervous system also causes the adrenal glands to release the hormone **adrenalin** into the bloodstream. This hormone closely resembles the chemical transmitter substance noradrenalin and produces all the same effects as the sympathetic nervous system but sustains them for a longer time. Thus nerve stimulation provides the quick response needed in emergencies of 'fight or flight' but the hormone response, although slower in onset, is long lasting and holds the body in readiness for a longer period until the crisis is over.

Revision questions

1 During which period of heartbeat does blood fill the atria?

2 Why is it essential for the wave of excitation to spread to all parts of the ventricles at great speed?

3 (a) Which part of the autonomic nervous system is most active in a person relaxing after a meal?
(b) Which of this person's arterioles would be constricted?

4 During exercise the cardiac output (volume of blood leaving the heart) can increase from 5 to 30 litres per minute. In what two ways can this be achieved?

5 If stimulation of the parasympathetic nervous system causes flushing of the skin, why is adrenalin used in first aid to stop bleeding in open wounds?

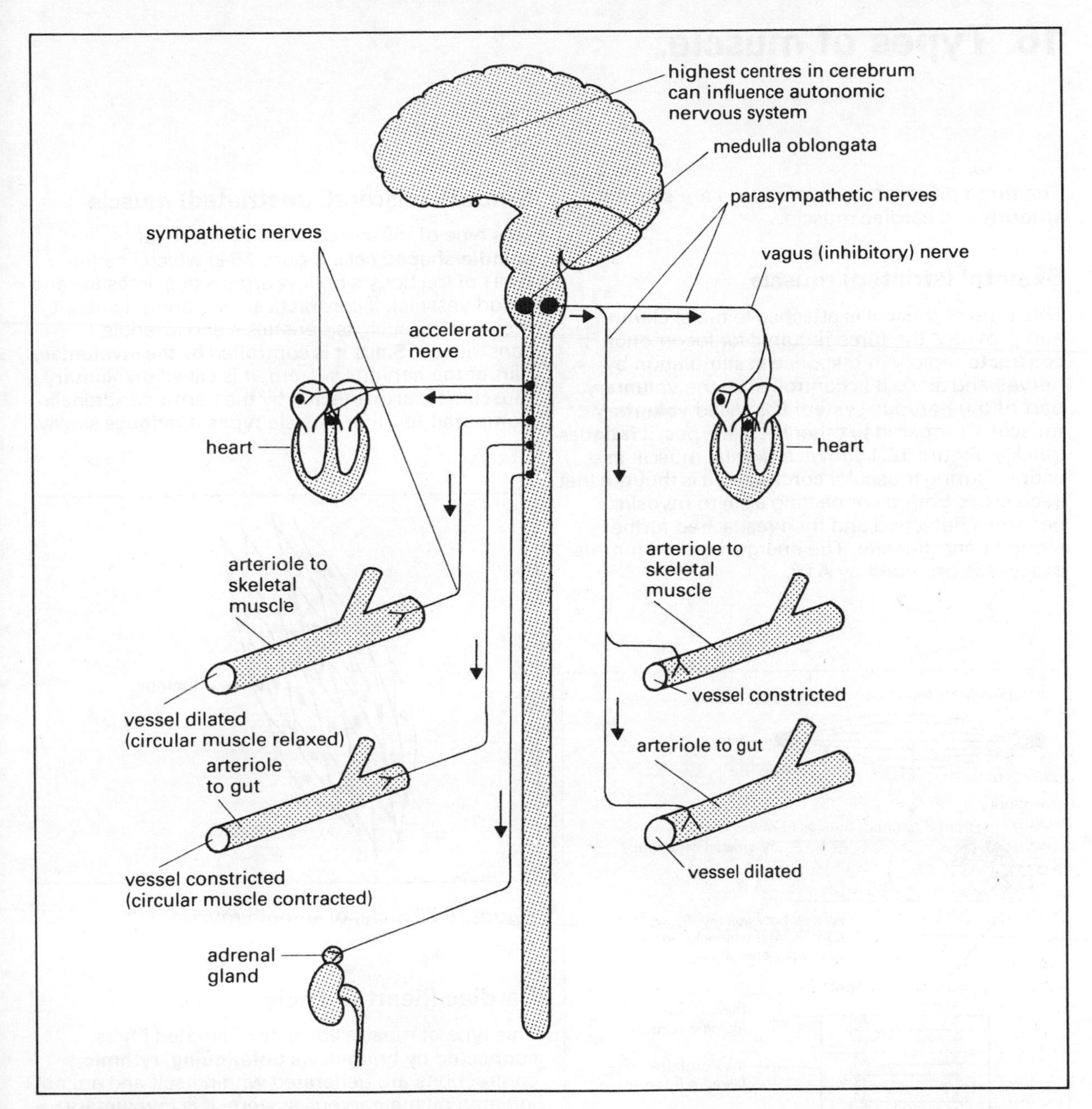

Figure 15.3 Autonomic nervous system

16 Types of muscle

The three different types of muscle are **skeletal, smooth** and **cardiac** muscle.

Skeletal (striated) muscle

This type of muscle is attached to the skeleton and provides the force required for locomotion. It **contracts** rapidly in response to stimulation by nerves and since it is controlled by the voluntary part of the nervous system it is called **voluntary** muscle. Compared to other muscle types, it **fatigues** quickly. Figure 16.1 shows a skeletal muscle in action. During muscular contraction it is thought that each cross bridge connecting actin to myosin becomes detached and then reattached further along at another site. The energy required for this process is provided by ATP.

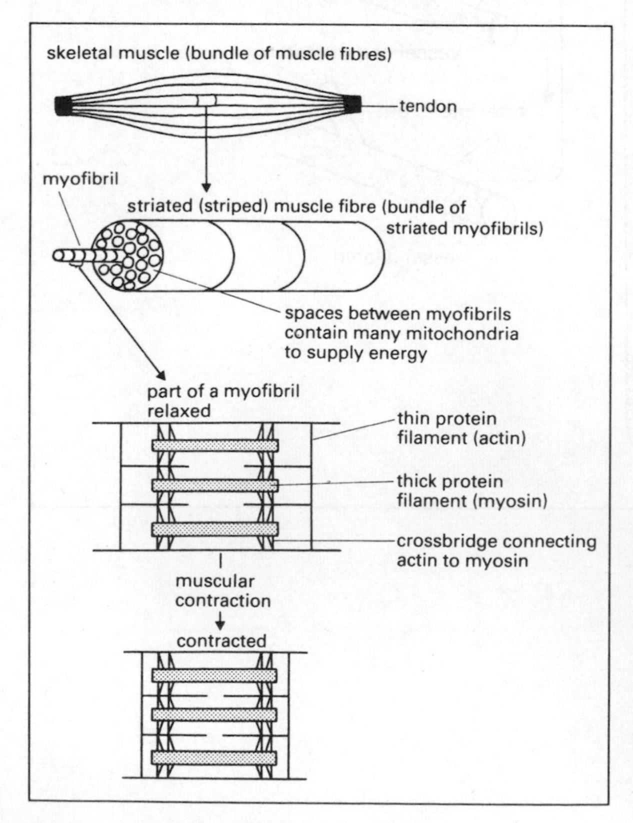

Figure 16.1 Skeletal muscle in action

Smooth (visceral, unstriated) muscle

This type of muscle consists of sheets of **spindle-shaped** cells (figure 16.2) which line the walls of the body's hollow organs (e.g. intestine and blood vessels). It **contracts** slowly bringing about movements such as peristalsis and arteriole constriction. Since it is controlled by the involuntary part of the nervous system, it is called **involuntary** muscle. It is also affected by the hormone adrenalin. Compared to other muscle types it **fatigues** slowly.

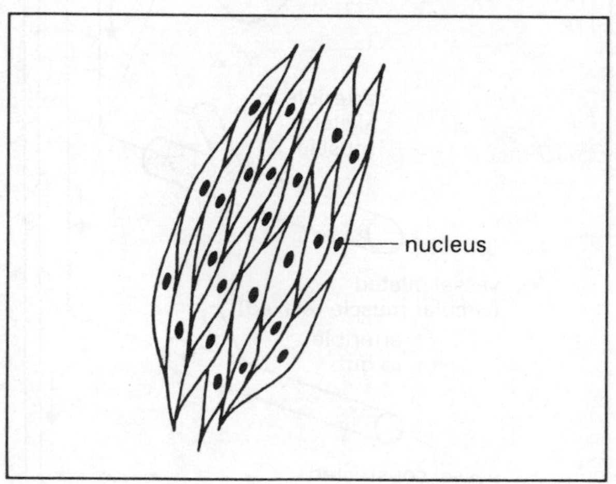

Figure 16.2 Sheet of smooth muscle cells

Cardiac (heart) muscle

This type of muscle consists of striated fibres connected by bridges. Its **unfatiguing, rythmic** contractions are generated within itself and are not initiated by the nervous system. It is **involuntary** but the rate and force of its contraction are affected by nerves and adrenalin (see chapter 15).

Comparison of vertebrate and invertebrate limb muscles

Vertebrate limb muscles are attached to the outside of an **endoskeleton.** The effect of muscular contraction is shown in figure 16.3. Invertebrate limb muscles are attached to the inside of an **exoskeleton.** The effect of muscular contraction is shown in figure 16.4.

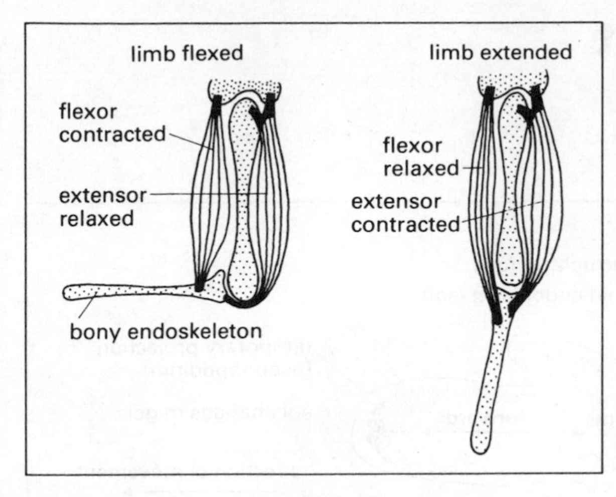

Figure 16.3 Vertebrate limb muscles (e.g. man)

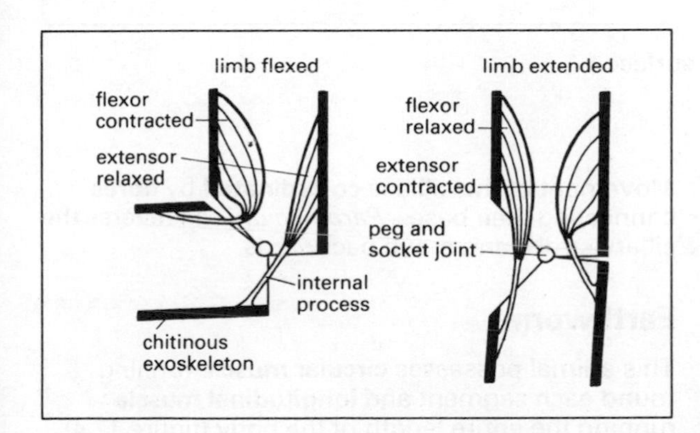

Figure 16.4 Invertebrate limb muscles (e.g. insect)

On contraction, each skeletal muscle brings about movement of one joint in one plane. Each muscle has to be restored to its original position by its **antagonist**. Thus successful locomotion in both of these animal types is effected by each limb possessing several joints which flex in different planes with each joint being activated by at least one pair of antagonistic muscles.

Revision questions

1 Figure 16.5 shows one section of a myofibril (cross bridges between protein filaments have not been included). (a) Which muscle type consists of such myofibrils? (b) Name protein filaments X and

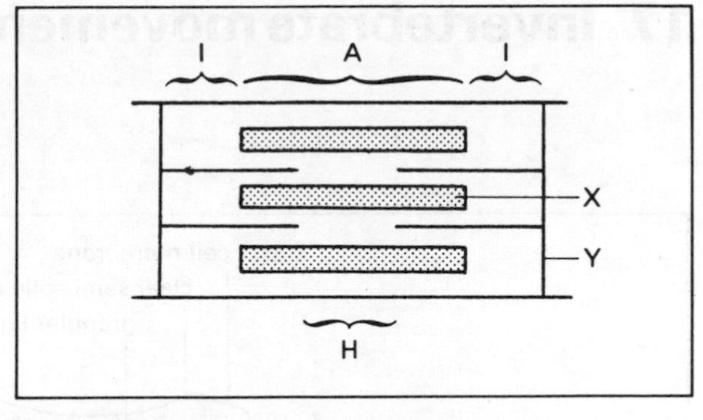

Figure 16.5

Y. (c) On contraction, what will happen to the lengths of bands I, A and H?

2 (a) Which muscle type is striated yet does not fatigue?
(b) Which muscle type is found in the wall of the mammalian uterus?

3 Contraction is a muscle's only positive action. How, then is the opposite movement of a limb possible?

4 (a) Name the type of skeleton shown in the diagram of a crab's leg.
(b) What effect would result from contraction of muscle X?
(c) In what way would the arrangement of the muscles in a mouse's leg differ from the crab (see figure 16.6).

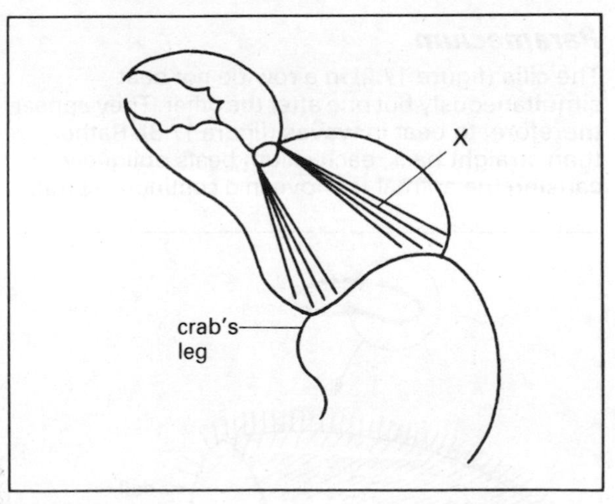

Figure 16.6

17 Invertebrate movement

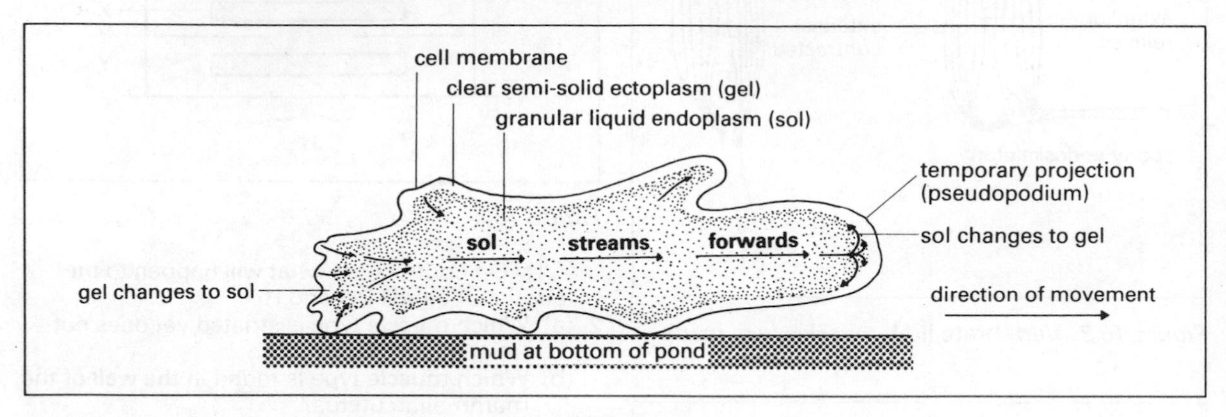

Figure 17.1 Amoeba moving in contact with a solid surface

Amoeba

By continuously building, ahead of itself, a tube through which a **liquid core** flows, *Amoeba* (figure 17.1) progresses in a slow and irregular manner, often putting out **pseudopodia** in directions opposite to previous advances.

Paramecium

The **cilia** (figure 17.2) in a row do not beat simultaneously but one after the other. They appear, therefore, to beat in **waves** (figure 17.3). Rather than straight back, each cilium beats obliquely causing the animal to move in a continous **spiral.**

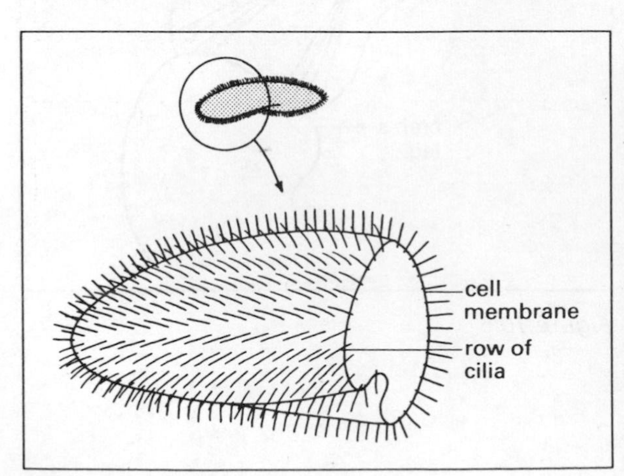

Figure 17.2 Paramecium

Movement of the cillia is **co-ordinated** by **fibres** connecting their bases. *Paramecium* can reverse the ciliary stroke and move backwards.

Earthworm

This animal possesses **circular muscle** running round each segment and **longitudinal muscle** running the entire length of the body (figure 17.4). The **coelom** (body cavity) contains liquid under pressure. It acts, therefore, as a **hydrostatic skeleton,** maintaining body shape.

Locomotion

As shown in figure 17.5 the anterior region of the worm advances by being squeezed as far forward as possible by contraction of the circular muscles (and relaxation of the longitudinal muscles).

Next the most anterior segments become short and fat on contraction of the longitudinal muscles (and relaxation of the circulars). The bristles **(cheatae)** now come out and grip the soil to anchor this region of the worm while the part behind moves along.

The anterior region now begins to progress forward again. Thus the worm moves forward by this alternate contraction and relaxation of antagonistic muscles occurring all along the body.

Insect

All of an insect's **locomotor** muscles act across joints and are attached to the inside of either the

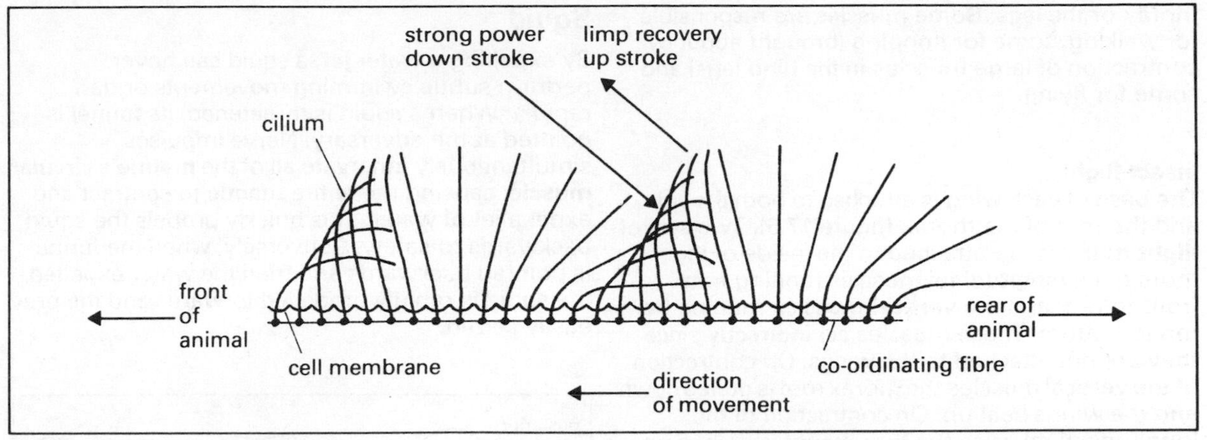

Figure 17.3 Ciliary action

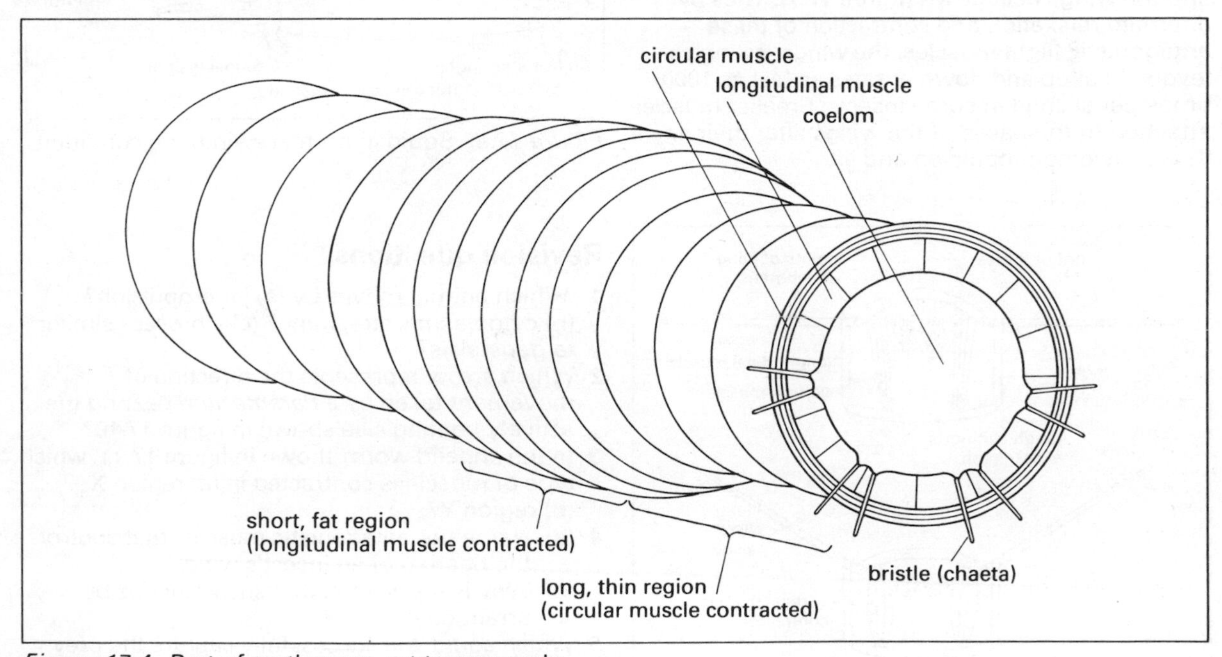

Figure 17.4 Part of earthworm cut transversely

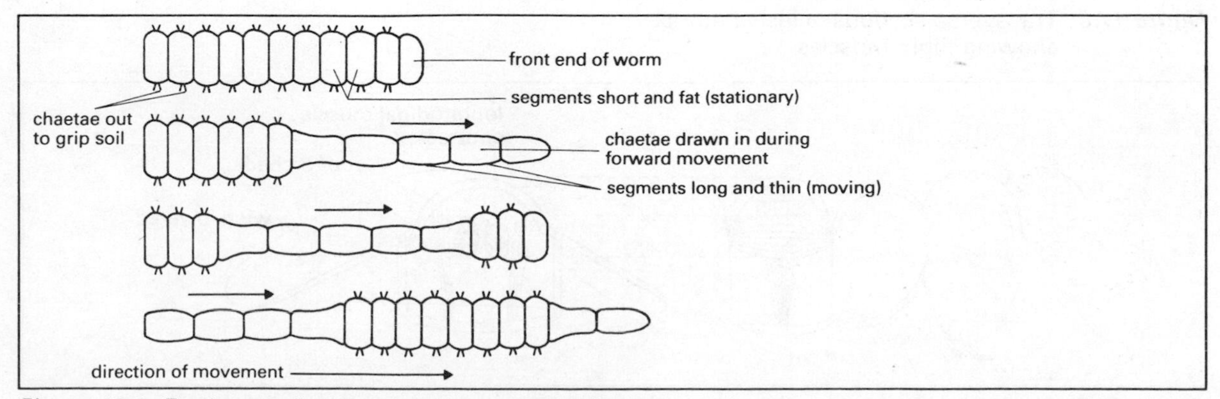

Figure 17.5 Earthworm locomotion viewed from above

thorax or the legs. Some muscles are responsible for walking, some for hopping (brought about by contraction of large muscles in the hind legs) and some for flying.

Insect flight

The base of each wing is attached to both the side and the roof of the thorax (figure 17.6). Two sets of **flight muscles** are attached to the inside of the thorax, the **longitudinal** muscles running from front to rear and the **vertical** muscles running from top to bottom. These muscles act **indirectly** since they are not attached to the wings. On contraction of the vertical muscles the thorax roof is pulled down and the wings beat up. On contraction of the longitudinal muscles the thorax roof is arched up and the wings beat down (figure 17.7). Thus by alternate relaxation and contraction of these antagonistic flight muscles, the wings, acting as levers, beat up and down at rates as fast as 1000 times per second in some insects. Smaller muscles attached to the bases of the wings alter their angle thus providing propulsion and lift.

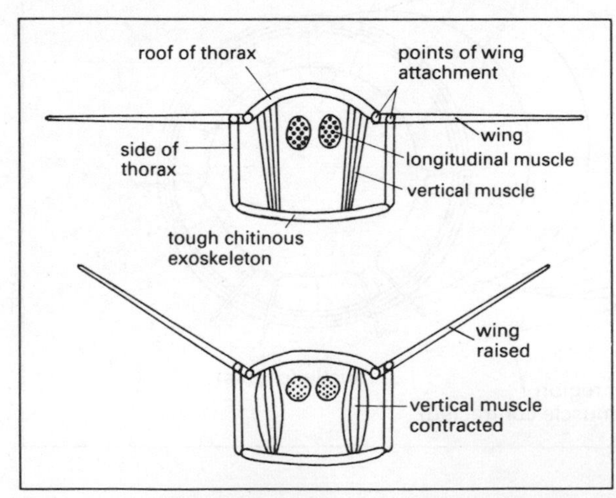

Figure 17.6 Transverse sections of insect thorax showing flight muscles

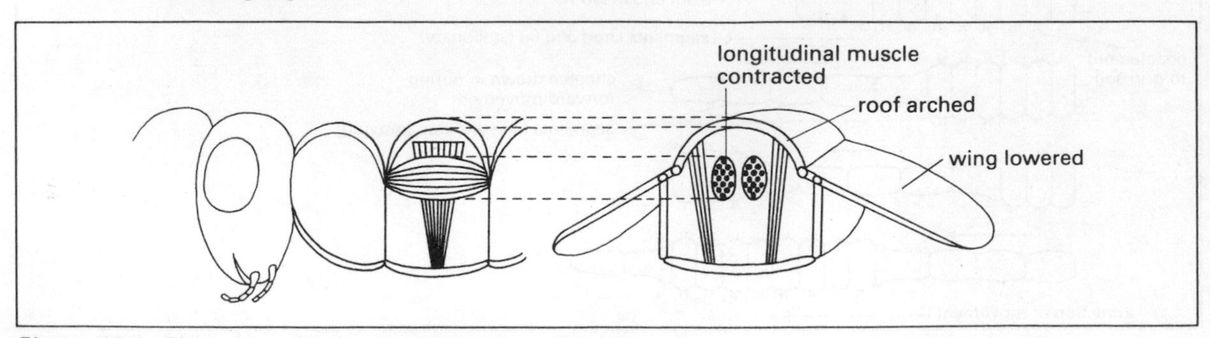

Figure 17.7 Side view and transverse section of thorax

Squid

By expelling a water **jet,** a squid can hover, perform subtle swimming movements or dart rapidly. When a squid is threatened, its **funnel** is pointed at the adversary. Nerve impulses simultaneously innervate all of the **mantle's circular muscle,** causing the entire mantle to contract and expel a jet of water. This quickly propels the squid backwards to safety. Conversely, when the funnel is pointed backwards and then the water expelled, the squid is rapidly propelled forwards and the prey easily seized.

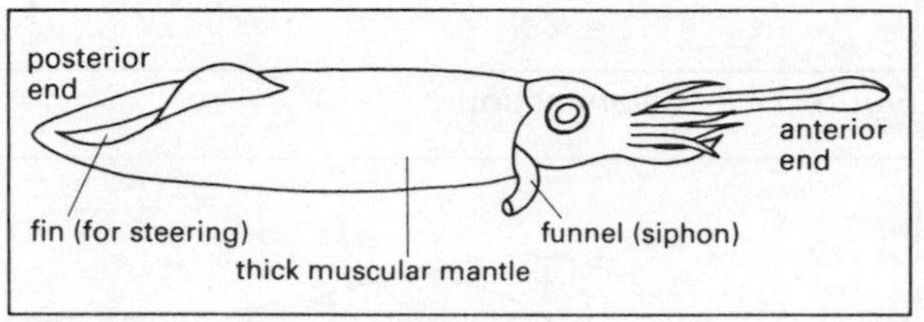

Figure 17.8 Squid in normal swimming condition

Revision questions

1 Which animal moves by (a) jet propulsion? (b) cytoplasmic streaming? (c) a process similar to peristalsis?
2 Which arrow represents the direction of movement taken by a *Paramecium* bearing the actively beating cilia shown in figure 17.10?
3 In the annelid worm shown in figure 17.11, which type of muscle is contracted in (a) region X (b) region Y?
4 (a) Name the antagonistic muscles that control the beating of an insect's wing.
(b) Why is it essential that such muscles be arranged in pairs?
5 Which squid will successfully capture the prey when its mantle contracts (see figure 17.12)?

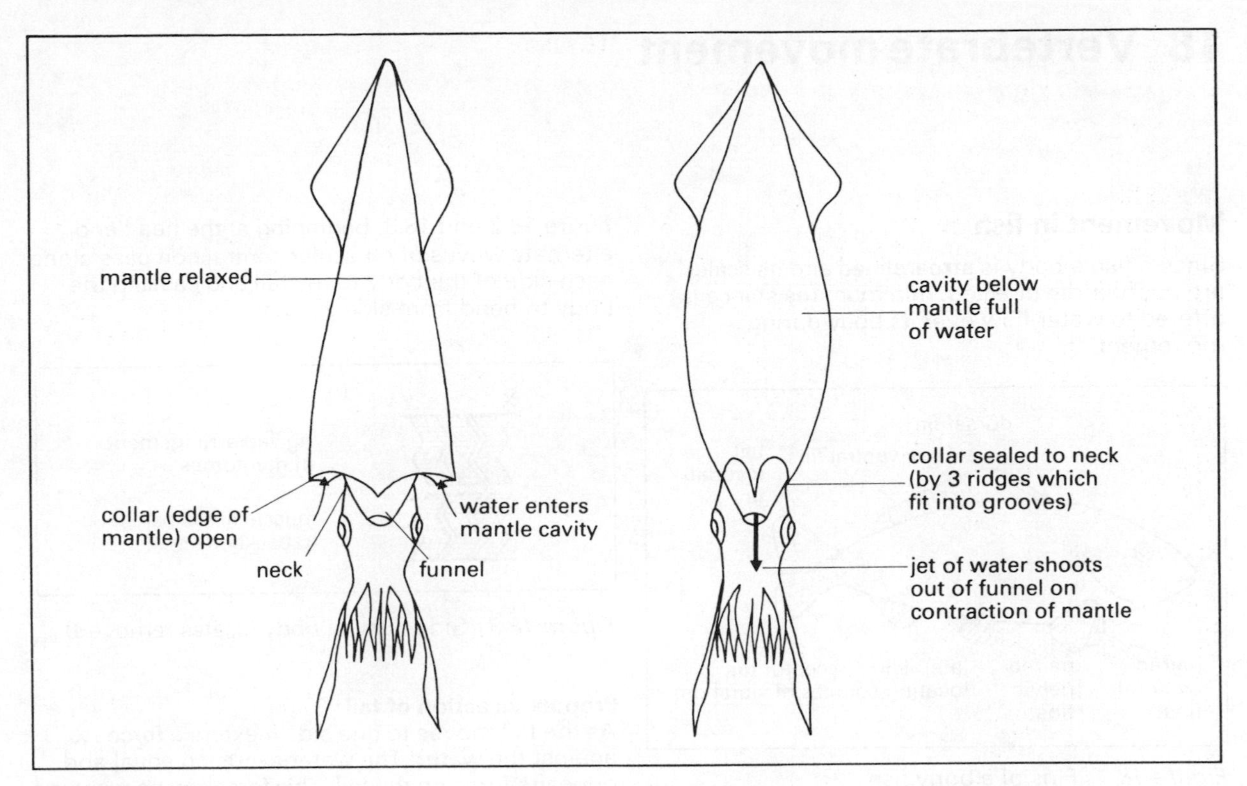

Figure 17.9 Ventral views of squid

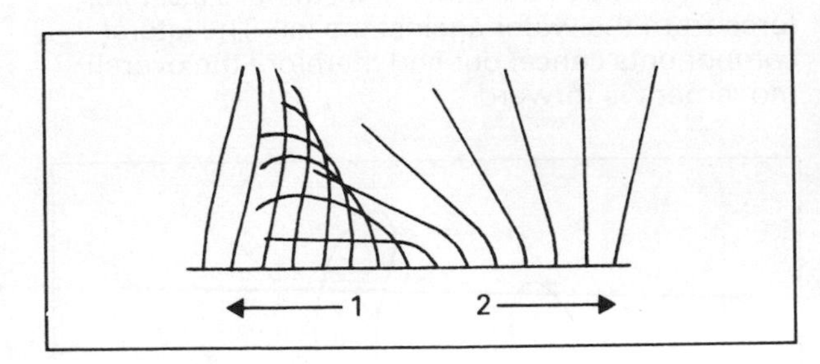

Figure 17.10

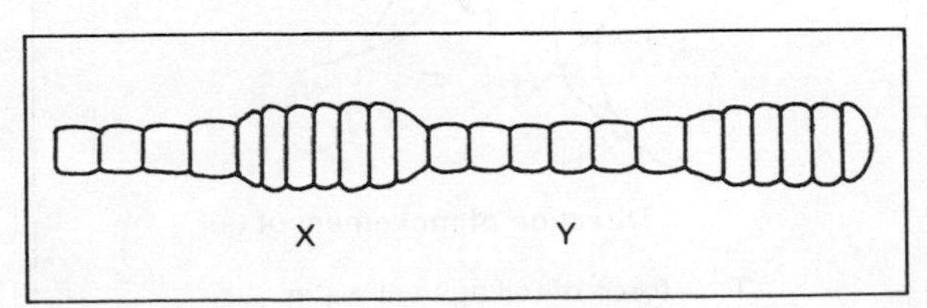

Figure 17.11

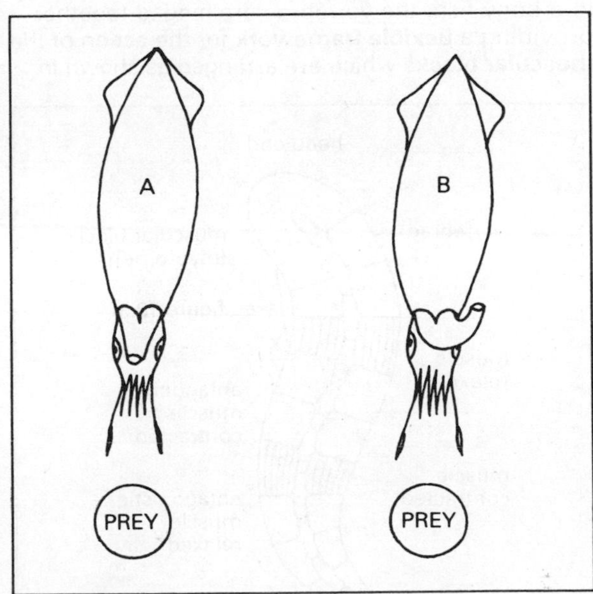

Figure 17.12

18 Vertebrate movement

Movement in fish

Since a fish's body is **streamlined** and its scales are backwardly directed, minimum resistance is offered to water flow over its body during movement.

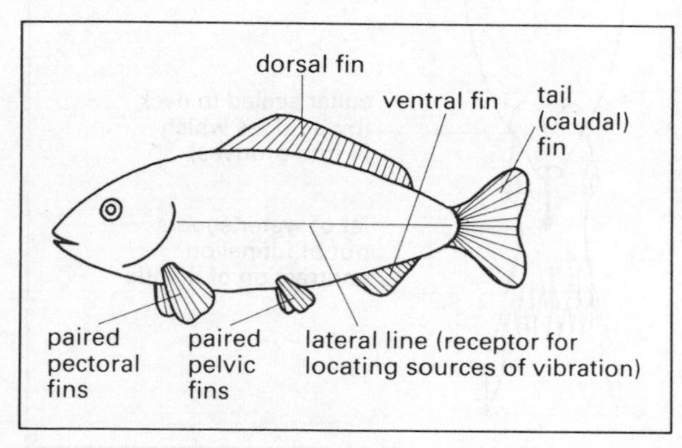

Figure 18.1 Fins of a bony fish

Use of muscles

In a bony fish, the vertebrae are hinged together providing a **flexible framework** for the action of the **muscular blocks** which are arranged as shown in figure 18.2 and 18.3. Beginning at the head end, alternate waves of muscular contraction pass along each side of the body to the tail end causing the body to bend from side to side.

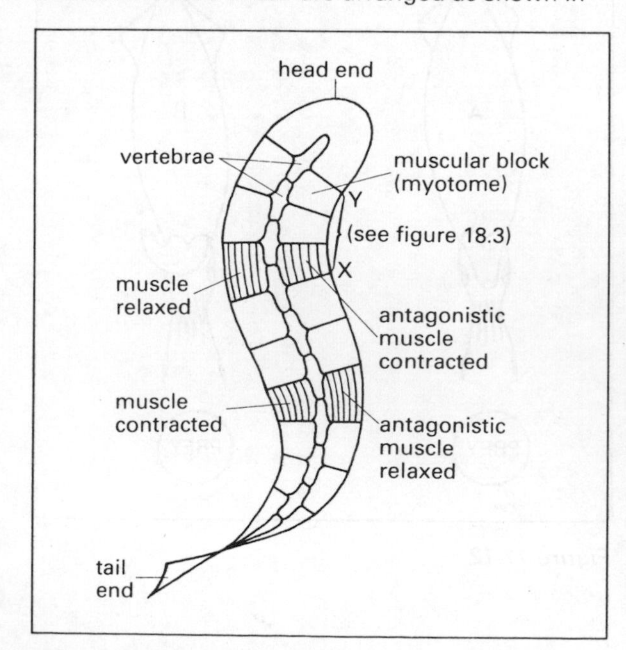

Figure 18.2 Dorsal view of muscle arrangement

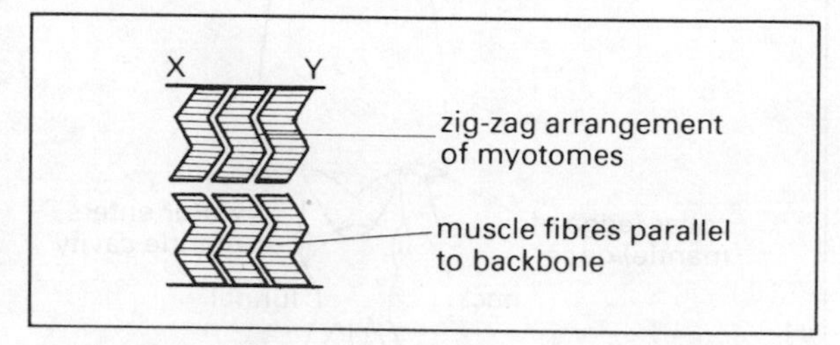

Figure 18.3 Side view of body (scales removed)

Propulsive action of tail

As the tail moves to one side it exters a **force** against the water. The water exerts an equal and opposite force on the tail. This force can be resolved into **forward** and **lateral** forces (figure 18.4). When the tail moves to the other side, there is a similar force from the water against the tail. The lateral components cancel out and therefore the overall movement is forward.

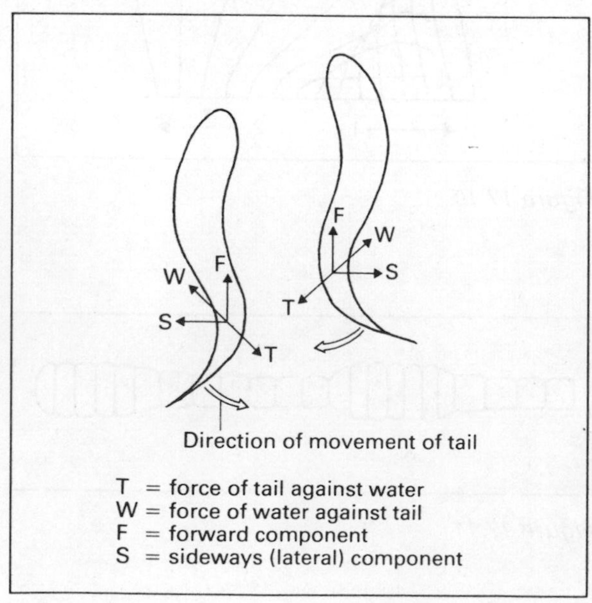

Figure 18.4 Propulsive action of tail

Use of fins
The tail fin increases the **active surface area** of the tail for forward propulsion. The other fins have the following functions.

Changing direction
A change in direction is brought about by the paired fins being held out at a certain angle.

Maintaining stability
A fish may be subjected to three types of **displacement** (figure 18.5). **Yawing** and **rolling** are controlled by the unpaired (dorsal and ventral) fins. **Pitching** is controlled by the paired (pectoral and pelvic) fins.

Braking
This is effected by the paired fins being held out at the sides therefore offering resistance to the flow of water over the body.

Buoyancy
Most bony fish possess a **swim bladder,** an air-filled sac opening into the back of the throat. The volume of gas in it can be varied allowing the fish to adjust its buoyancy and swim at different depths.

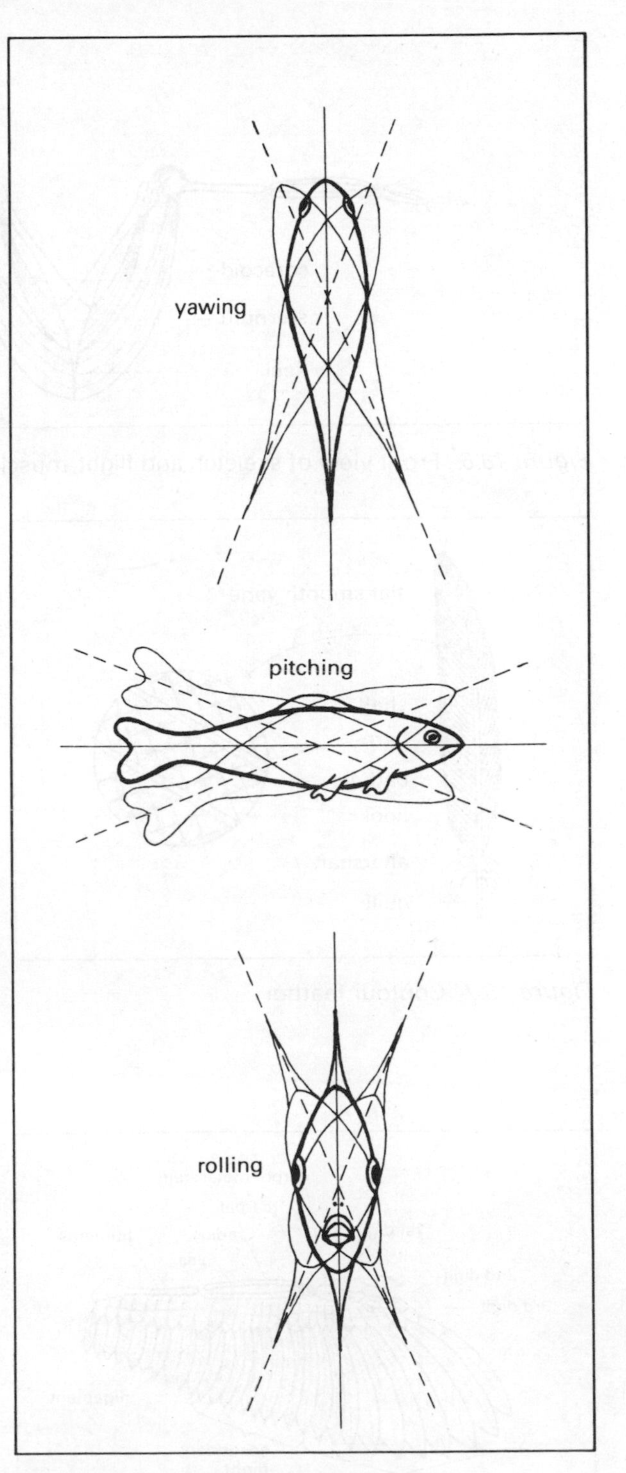

Figure 18.5 Types of displacement

Structural adaptations of birds for flight

Skeleton
Many bones contain **air spaces** and therefore are light in weight. A reduction in number of bones (e.g. wing has only three digits) also contributes to the lightness of body weight needed for flight. Some bones are **fused** together thus increasing their strength (e.g. wrist bones are fused together to form the strong **carpo-metacarpus** for attachment of flight feathers). The **sternum** bears a **keel-shaped** extension for attachment of flight muscles (figure 18.6).

Muscles
Birds possess powerful **pectoral muscles** (figure 18.6) for flight. On contraction, the **pectoralis major** brings about lowering of the wing (i.e. the powerful downstroke). Contraction of its smaller antagonistic partner, the **pectoralis minor,** raises the wing (i.e. weaker recovery stroke). The tendon of the pectoralis minor, by passing through a hole between the shoulder bones and over the deltoid proces, acts as a pulley giving the leverage required to raise the wing. In birds that fly, the muscles are red due to the presence of large quantities of **respiratory pigments** necessary to supply oxygen for muscular contraction.

Feathers
The structure of a **contour** feather is shown in figure 18.7. During **preening** disturbed **barbs** are hooked back together by their **barbules.** Oil from the preen gland waterproofs the feathers. Contour feathers on the body overlap providing a smooth surface and minimising air resistance. The arrangement of flight feathers is shown in figure

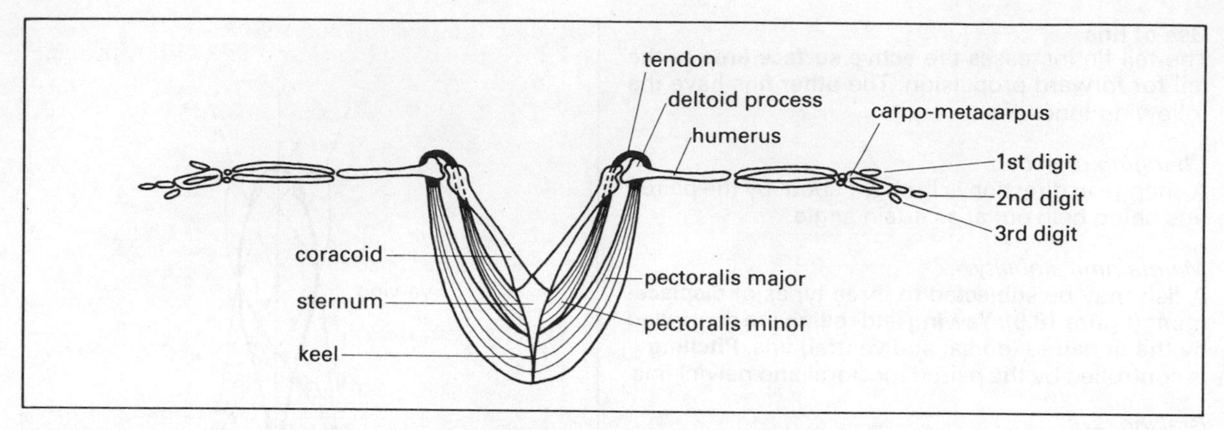

Figure 18.6 Front view of skeleton and flight muscles

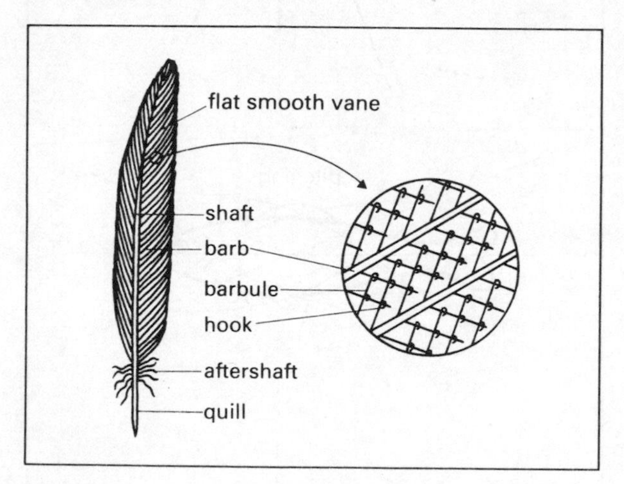

Figure 18.7 Contour feather

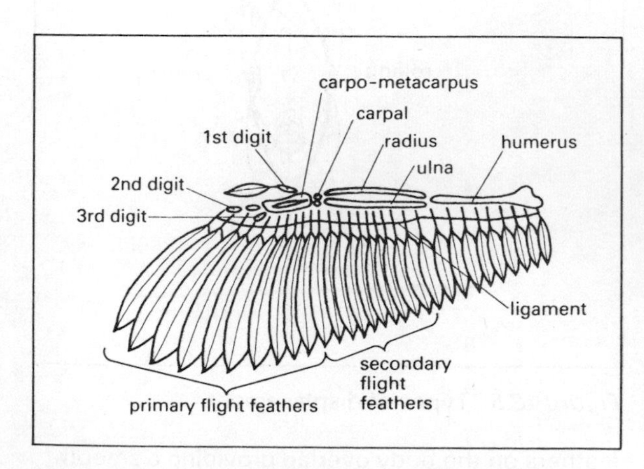

Figure 18.8 Arrangement of bones and flight feathers in wing

18.8. On the downstroke the overlapping feathers provide a large surface acting down on air and gaining **upthrust** (figure 18.9). This lift is not lost during the upstroke because the feathers are opened reducing resistance to air. **Down** feathers (figure 18.10) are found under the contour feathers where they trap air for insulation.

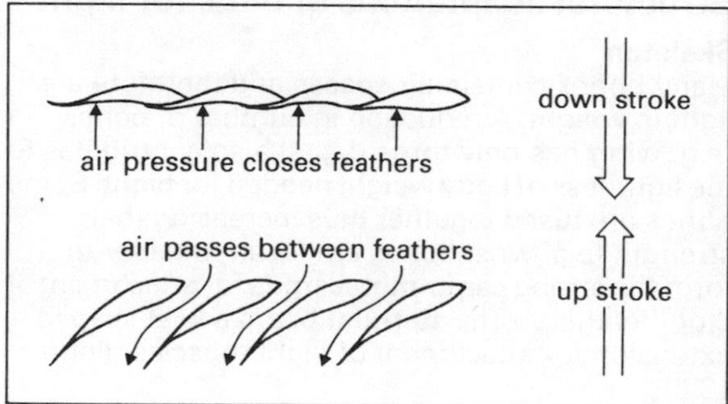

Figure 18.9 Arrangement of wing feathers during flight

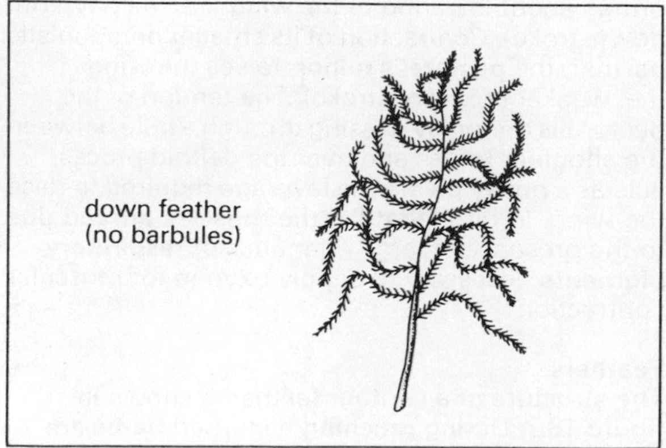

Figure 18.10 Down feather (no barbules)

Air sacs attached to lungs
These structures (figure 18.11) help to make the body lighter and allow fresh air to pass through the lungs during **expiration** thus providing extra **oxygen** for energy.

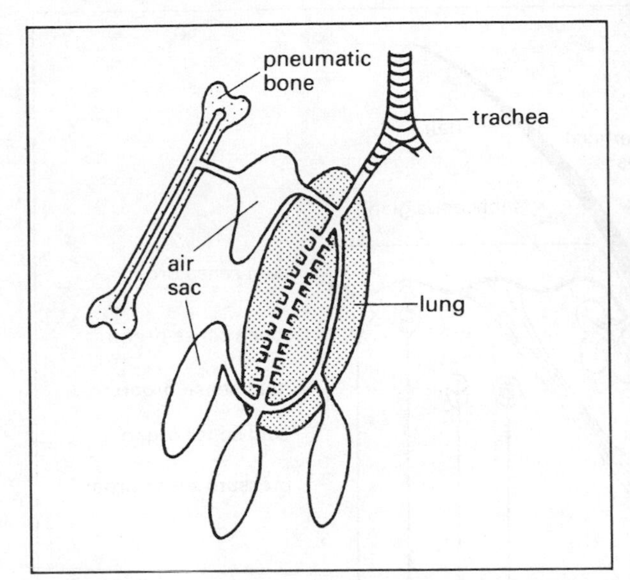

Figure 18.11 Air sacs attached to lungs

Large four-chambered heart and high pulse rate
These give increased **blood supply** to support the high metabolic rate required for flight.

Large optic lobes and cerebellum
These regions of the brain provide excellent **sight** and **balance.**

Fat reserve
This acts as an **insulating layer** and provides energy during flight.

Revision questions

1. Which myotome is antagonistic to M (see figure 18.12)?
2. Which of a fish's fins are involved in (a) turning (b) counteracting pitch?
3. Which muscle raises the wing of a bird? To which bones are the tendons of this muscle attached?
4. Which two structural modifications provide extra oxygen in birds that fly?
5. The barbs of an emu's feathers do not possess hooks. Explain why an emu cannot fly.

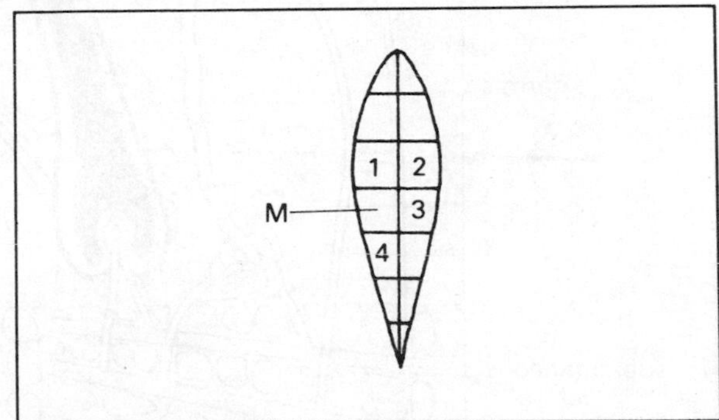

Figure 18.12

19 Skin and temperature regulation

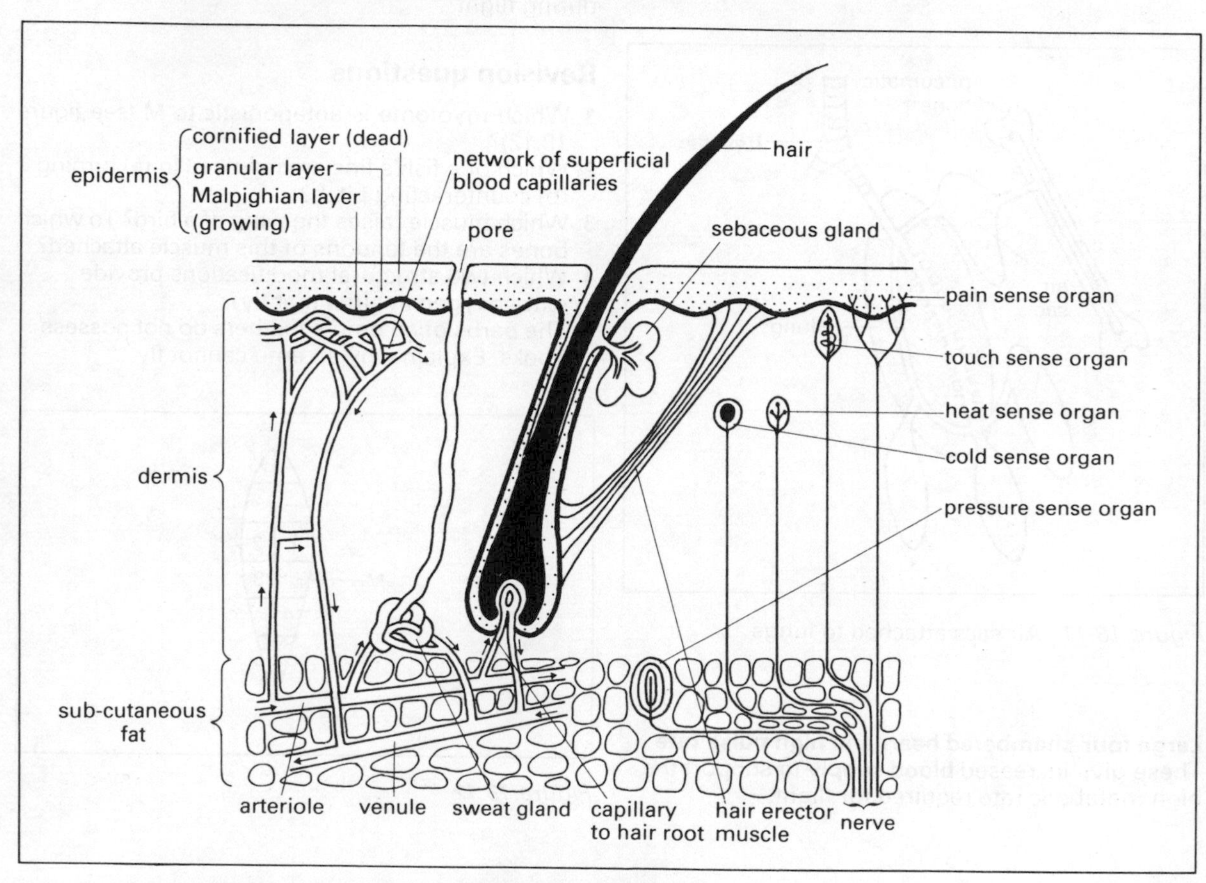

Figure 19.1 Structure of mammalian skin

Functions of the skin

Protection

The skin protects underlying tissues and organs from damage due to **friction, desiccation,** entry of **micro-organisms** (some of which could cause disease) and damage by **ultra violet (UV) radiation** in sunlight. The dark pigment **melanin** made by the Malpighian layer (figure 19.1) in response to UV light prevents UV rays penetrating and damaging vital organs. Oily sebum from the sebaceous glands lubricates and waterproofs skin and hair.

Sensitivity

Sensory receptors (figure 19.1) detect **touch, pain, temperature** changes and **pressure.** Sensory receptors are more numerous in some areas of the body than others. Lips and fingertips are especially sensitive.

Excretion

Excess salts and a small amount of urea are excreted in **sweat** from the sweat glands (figure 9.1).

Regulation of body temperature

Overheating of the body is corrected in two ways.

Increase in rate of sweating
Heat energy from the body is used to convert the water in **sweat** to **water vapour** therefore bringing about a lowering of body temperature. Some

mammals (e.g. dog) do not possess sweat glands and lose heat from the tongue by **panting.**

Vasodilation of arterioles in the dermis (figure 19.2)
Excess heat is easily lost by **radiation** from the superficial capillaries when they become filled with blood.

Overcooling of the body is corrected in four ways.

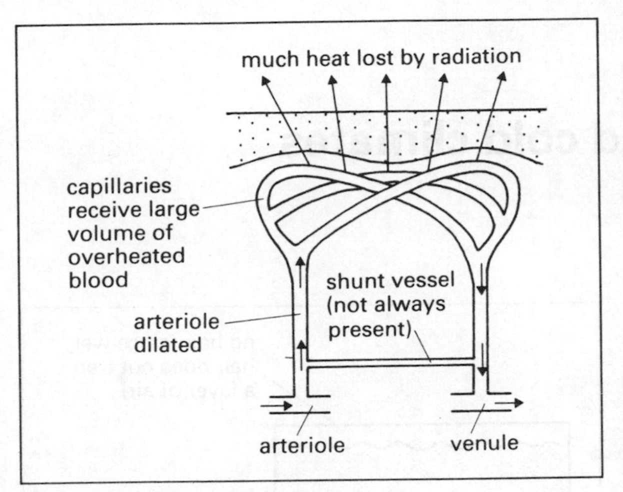

Figure 19.2 Vasodilation

Decrease in rate of sweating

Vasoconstriction of arterioles in the dermis (figure 19.3)
Minimum blood flows to the superficial capillaries and therefore little heat is lost by radiation. On some occasions blood by-passes the surface by flowing along a **shunt** vessel.

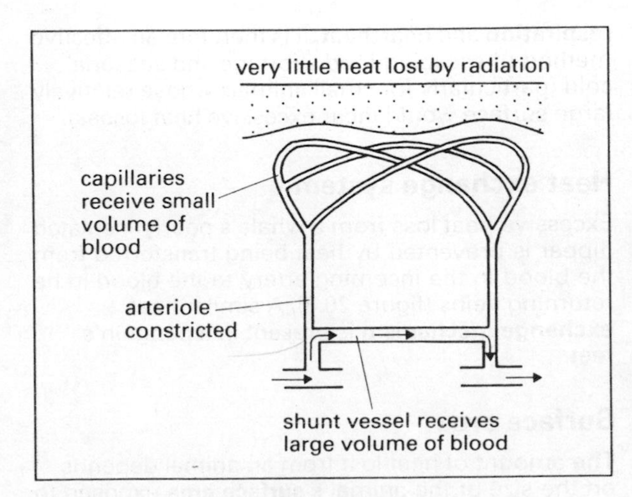

Figure 19.3 Vasoconstriction

Contraction of hair erector muscles
This results in hairs being raised to an almost vertical position (figure 19.4). A wide layer of **air** (a **poor conductor** of heat) is now trapped between the animal's body and the external environment. This layer of insulation reduces heat loss.

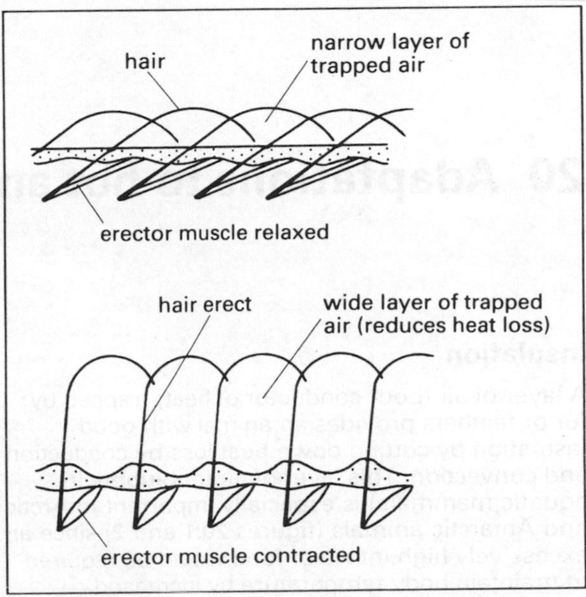

Figure 19.4 Action of hair erector muscles

Shivering (muscular contractions) and increase in metabolic rate
Both of these processes generate heat energy and therefore help to return the overcooled mammal's body temperature back to normal.

All mechanisms of temperature regulation are controlled by the **hypothalamus** in the brain. Animals which are able to maintain constant body temperature despite changes in the external environment are **homoiothermic** (birds and mammals). Animals whose body temperture varies according to the temperature of the external environment are **poikilothermic** (invertebrates, fish, amphibians and reptiles). Some poikilothermic animals can, however, exert some control over their body temperature by **behavioural** means. For example, many lizards and snakes move into and out of sunlight as required.

Revision questions

1 In cold weather birds fluff out their feathers. (a) What structures in the skin bring about this response? (b) How do fluffed-out feathers help to regulate body temperature?

2 What substance is made by the skin to prevent UV rays entering and damaging the body's internal organs?

3 Which sensory receptors are present in the (a) epidermis (b) dermis and (c) sub-cutaneous fat of man's skin?

4 (a) Why is a sweat gland found to be in intimate contact with a network of blood capillaries?
(b) To make up for the absence of sweat glands, kangaroos lick their fur in hot weather. Explain how this brings about a lowering of body temperature.

20 Adaptations to hot and cold climates

Insulation

A layer of air (poor conductor of heat) trapped by fur or feathers provides an animal with good insulation by cutting down heat loss by conduction and convection. This layer of air (or **blubber** in aquatic mammals) is especially important to Arctic and Antarctic animals (figures 20.1 and 2) since an excessively high intake of food would be required to maintain body temperature by increased metabolism alone. A camel's coarse coat traps a layer of air which reduces both heat loss at night and heat gain during the day.

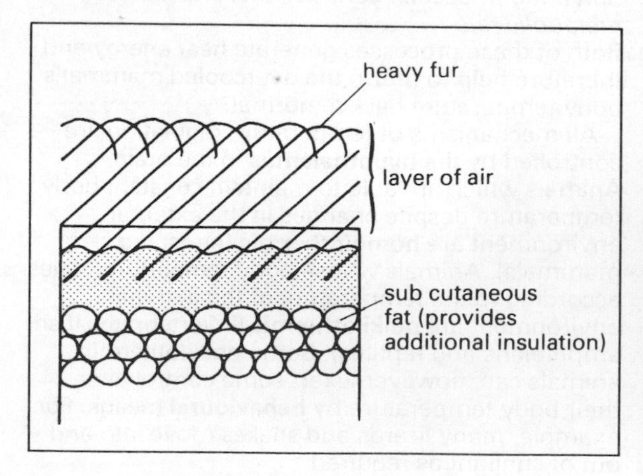

Figure 20.1 Skin of terrestrial Arctic mammal (e.g. husky dog)

Hibernation

Hibernation involves a profound drop in the animal's body temperature and rate of **metabolism**,

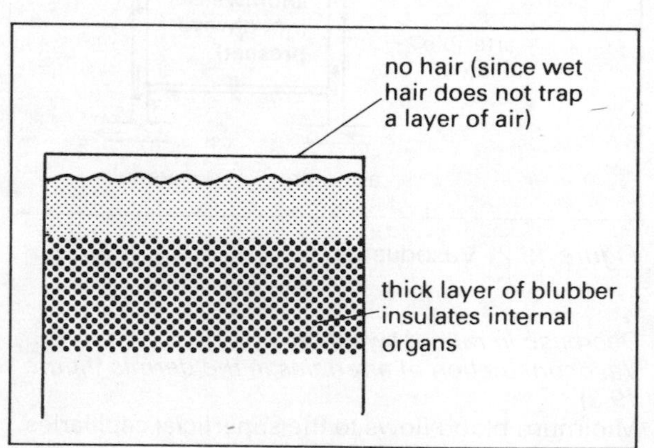

Figure 20.2 Skin of aquatic Arctic mammal (e.g. whale)

respiration and **heartbeat**. It is therefore an effective method of escaping food shortage and seasonal cold (particularly for small animals whose relatively large surface would incur excessive heat losses).

Heat exchange system

Excessive heat loss from a whale's poorly insulated flipper is prevented by heat being transferred from the blood in the incoming artery to the blood in he returning veins (figure 20.3). A similar **heat exchange mechanism** is present in a penguin's feet.

Surface area

The amount of heat lost from an animal depends on the size of the animal's **surface area** exposed to the environment. This varies in three ways.

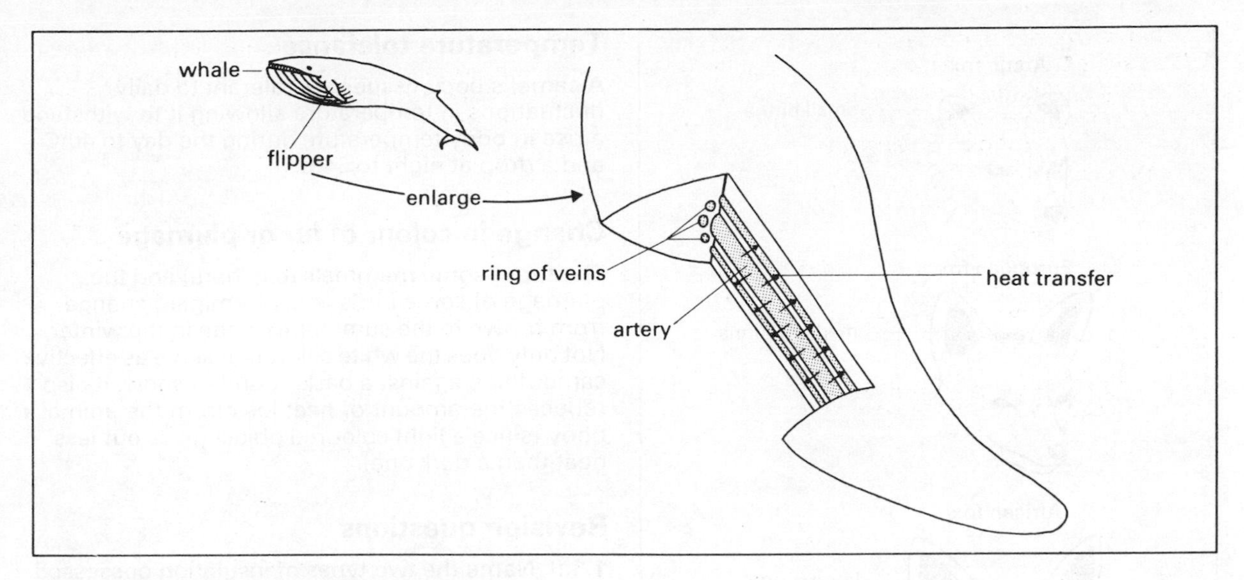

Figure 20.3 Heat exchange system

Breeding ground	Average body length
Greenland	320 mm
Norway	300 mm
United Kingdom	290 mm

Figure 20.4 Body size in puffins

Body size
Animals in cold climates are often larger in body size than related species in warm climates since a larger body has a relatively smaller surface area from which heat can be lost (figure 20.4).

Body shape
Two persons of equal mass and volume (figure 20.5) may differ in shape. A desert dwelling person is tall and thin with long limbs and a shallow body. The resultant larger surface area promotes heat loss. The reverse is true of a fat, thickset, rounded Eskimo.

Body extremities
A mammal's outer ear (pinna) is a thin layer of tissue containing many blood capillaries. An Artic fox (figure 20.6) possesses **small pinnae** and

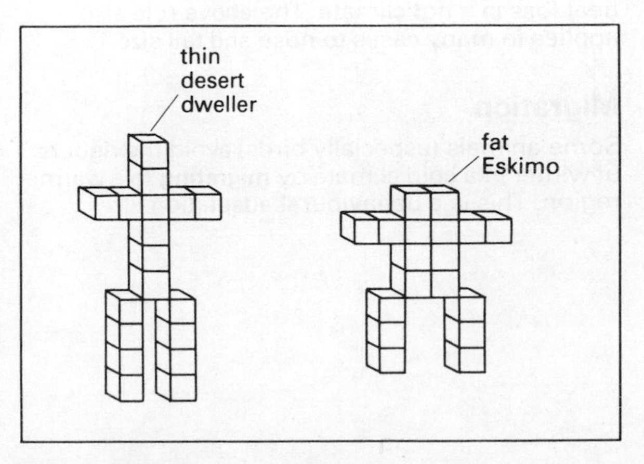

Figure 20.5 Body shape

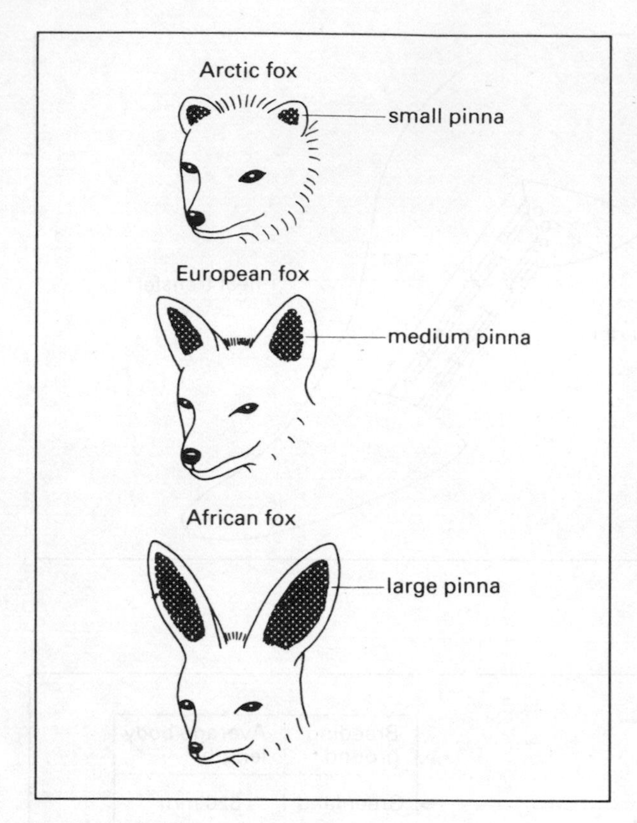

Figure 20.6 Pinna size in foxes

therefore a small surface area of blood capillaries is exposed to possible heat loss by radiation in a cold climate. The reverse is true of an African bat-eared fox whose **large pinnae** promote maximum heat loss in a hot climate. The above rule also applies in many cases to nose and tail size.

Migration

Some animals (especially birds) avoid the rigours of winter in a cold climate by **migrating** to a warmer region. This is a behavioural adaptation.

Temperature tolerance

A camel's body tissues are tolerant to daily fluctuations in temperature allowing it to withstand a rise in body temperature during the day to 40°C and a drop at night to 34°C.

Change in colour of fur or plumage

The fur of some mammals (e.g. hare) and the plumage of some birds (e.g. ptarmigan) change from brown in the summer to white in the winter. Not only does the white coloration serve as effective camouflage against a background of snow, it also reduces the amount of heat loss from the animal's body (since a light coloured object gives out less heat than a dark one).

Revision questions

1 (a) Name the two types of insulation possessed by a seal (see figure 20.7).
(b) Why does a seal require both types?

2 Explain why a humming bird (body length = 100 mm) hibernates every night.

3 (a) Why do polar bears tend to be rounder in shape than their brown relatives of similar mass?
(b) Compare the tails and pinnae possessed by those two types of bear.

4 Why do penguin chicks have thicker, fluffier coats than their parents?

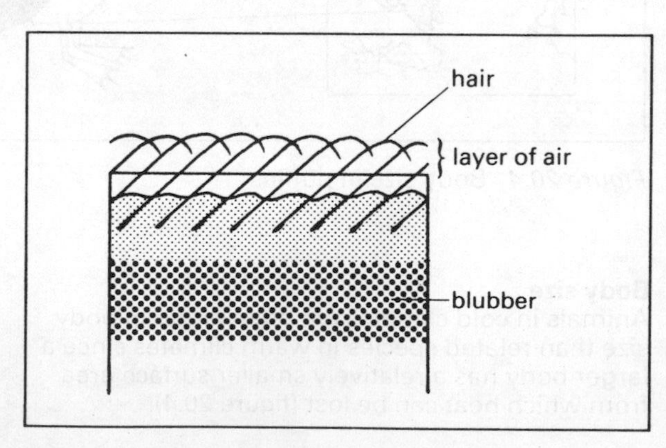

Figure 20.7

21 Animal embryology

The development of an animal embryo involves the following three initial stages.

Cleavage

During this first stage, repeated cell divisions occur, forming a **blastula.** Figure 21.1 shows cleavage as it occurs in both *Echinus* (sea urchin) and *Amphioxus.* There is no increase in size of the organism during this stage. As the number of cells increases, therefore, the size of each cell decreases. The size of each nucleus, however, remains constant.

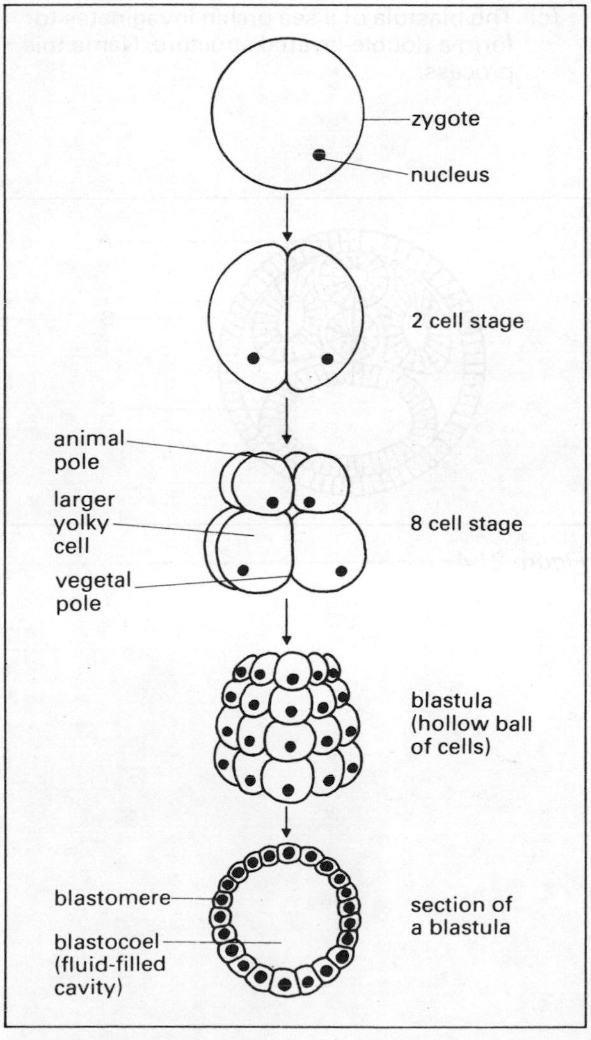

zygote
nucleus
2 cell stage
animal pole
larger yolky cell
vegetal pole
8 cell stage
blastula (hollow ball of cells)
blastomere
blastocoel (fluid-filled cavity)
section of a blastula

Figure 21.1 Cleavage

Gastrulation

This process of cell movement and arrangement into layers varies from one species to another. Figure 21.2 shows gastrulation as it occurs in *Echinus* and *Amphioxus.* The two-layered **gastrula** is formed as a result of the 'pushing in' of one side of the blastula.

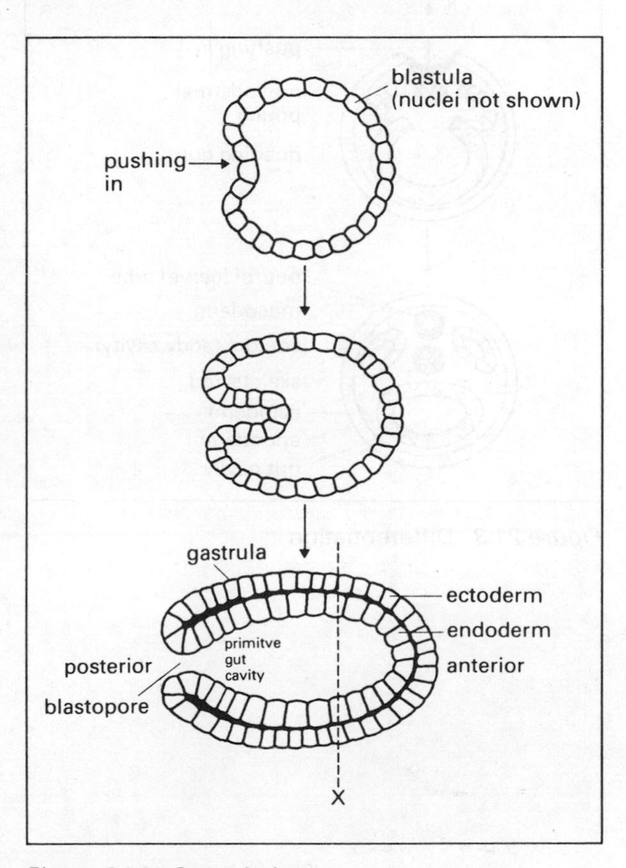

Figure 21.2 Gastrulation

Differentiation

This is the formation of specialised tissues and organs. Figure 21.3 shows early stages of differentiation in *Amphioxus.* Eventually all of the body's

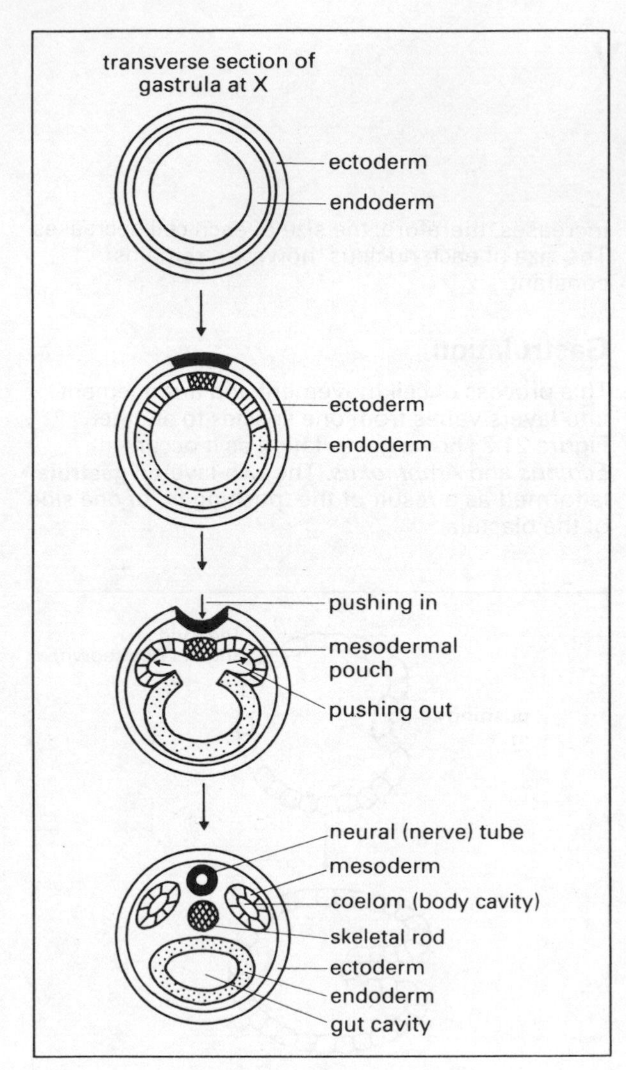

Figure 21.3 Differentiation

tissues and organs differentiate from the three basic layers shown. This is true of all **triploblastic** (three-layered) animals. For example, in man the **ectoderm** gives rise to skin and nervous tissue, the **mesoderm** to heart and muscles and the **endoderm** to the lining of the gut.

Revision questions

1 (a) Figure 21.4 shows an *Amphioxus* embryo undergoing differentiation. Explain the meaning of the term differentiation.
(b) Name structures A, B and C.
(c) From which of these is the animal's tail muscle derived?

2 (a) What name is given to the first mitotic division undergone by a zygote?
(b) Name the fluid-filled cavity found inside a blastula.
(c) The blastula of a sea urchin invaginates to form a double-layered structure. Name this process.

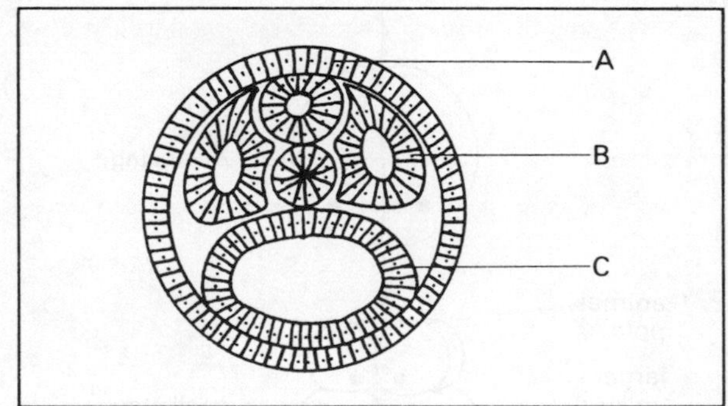

Figure 21.4

22 Apical meristems

A **meristem** is a group of **undifferentiated** plant cells which are capable of dividing repeatedly by **mitosis.** An **apical** meristem is found at a **root** and **shoot** tip (apex). Increase in length of a root or shoot depends on both the formation of new cells by the apical meristem and elongation of these new cells. Figures 22.1 and 22.2 show growth at a shoot and root apex respectively.

During the growth of a shoot or root, a new cell produced in a meristem by mitosis becomes **elongated, vacuolated** and finally **differentiated.** Differentiation is the process by which the unspecialised cell becomes altered and adapted to perform a special function as part of a permanent tissue.

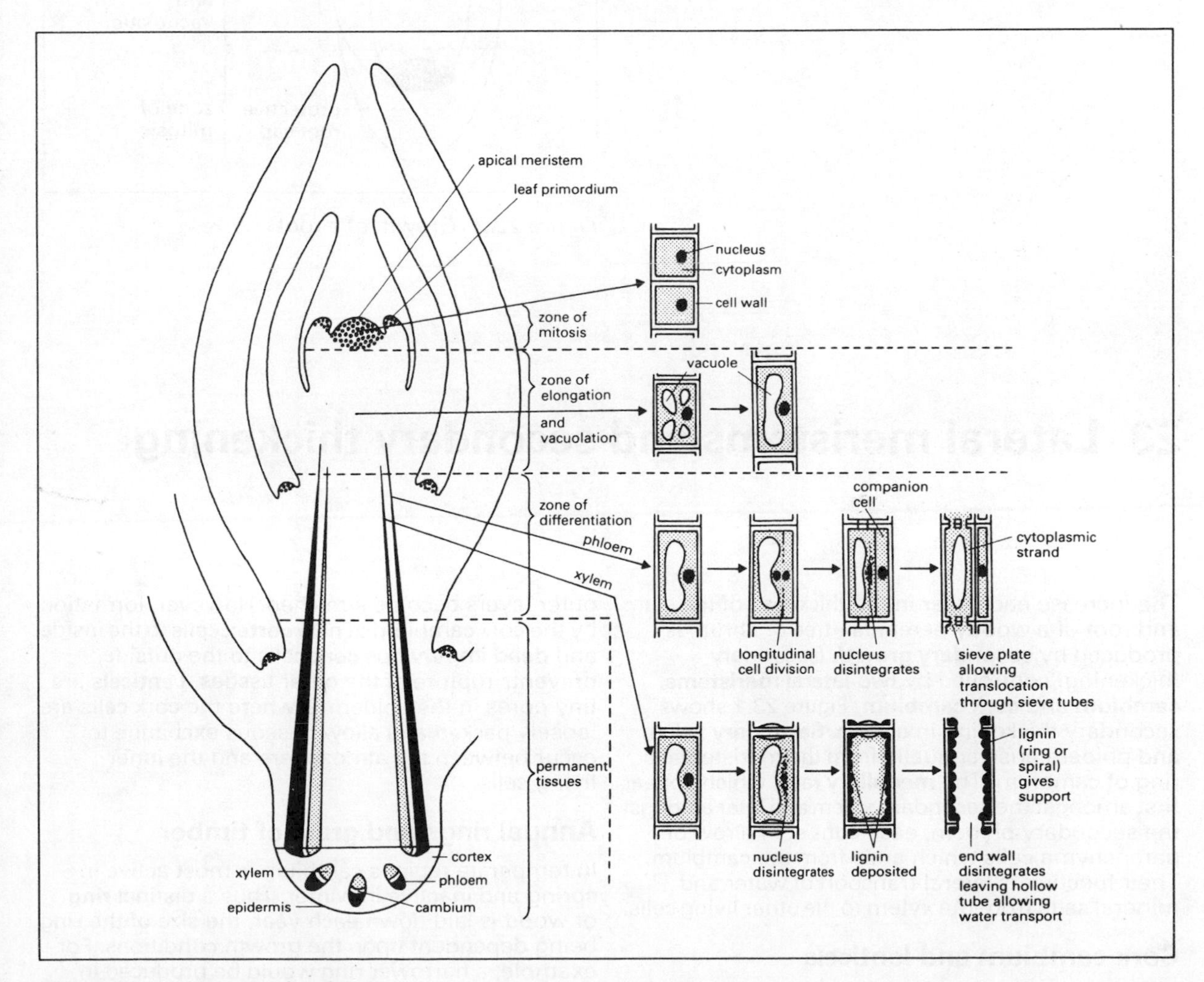

Figure 22.1 Growth of a shoot

Revision questions

1 State the function of meristematic cells.
2 Name two meristematic regions in a plant.
3 Which two processes produce downward growth of a root?
4 State three differences between a meristematic cell and a xylem vessel.

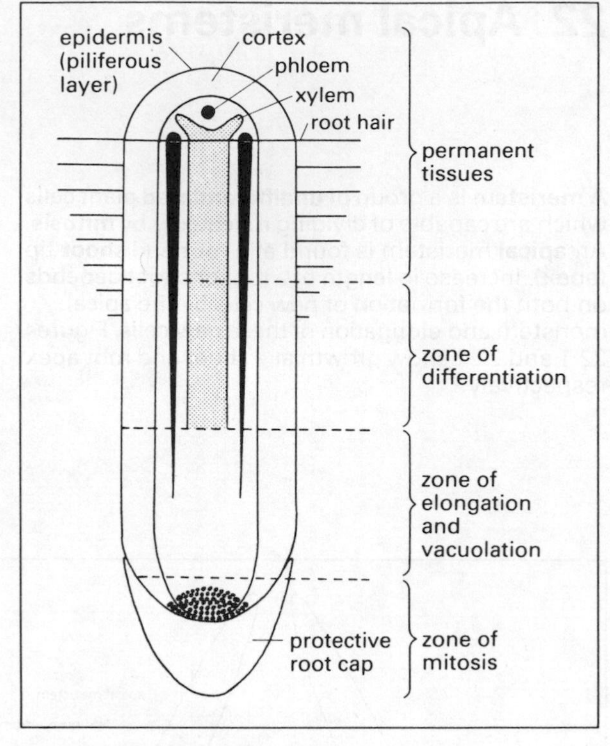

Figure 22.2 Growth of a root

23 Lateral meristems and secondary thickening

The increase each year in the thickness of the stem and root of a woody perennial (tree or shrub) is produced by **secondary growth** (secondary thickening) controlled by two **lateral meristems, cambium** and **cork cambium**. Figure 23.1 shows secondary thickening in a stem. **Secondary** xylem and phloem arise annually from the meristematic ring of cambium. The **medullary rays,** which appear first amongst the secondary xylem and later amongst the secondary phloem, each consist of a row of parenchyma cells which arise from the cambium. Their function is lateral transport of water and mineral salts from the xylem to the other living cells.

Cork cambium and lenticels

As the internal tissues increase in thickness, the outer layers become stretched. However, formation by the cork cambium of new **cortex** cells to the inside and **dead impervious cork** cells to the outside, prevents rupture of the outer tissues. **Lenticels** are tiny pores in the epidermis where the cork cells are loosely packed and allow gaseous exchange to occur between the atmosphere and the inner living cells.

Annual rings and grain of timber

In temperate regions cambium is most active in spring and inactive in winter. Thus a distinct **ring** of wood is laid down each year, the size of the ring being dependent upon the growth conditions. For example, a narrower ring would be produced in drought conditions. When timber is cut longitudi-

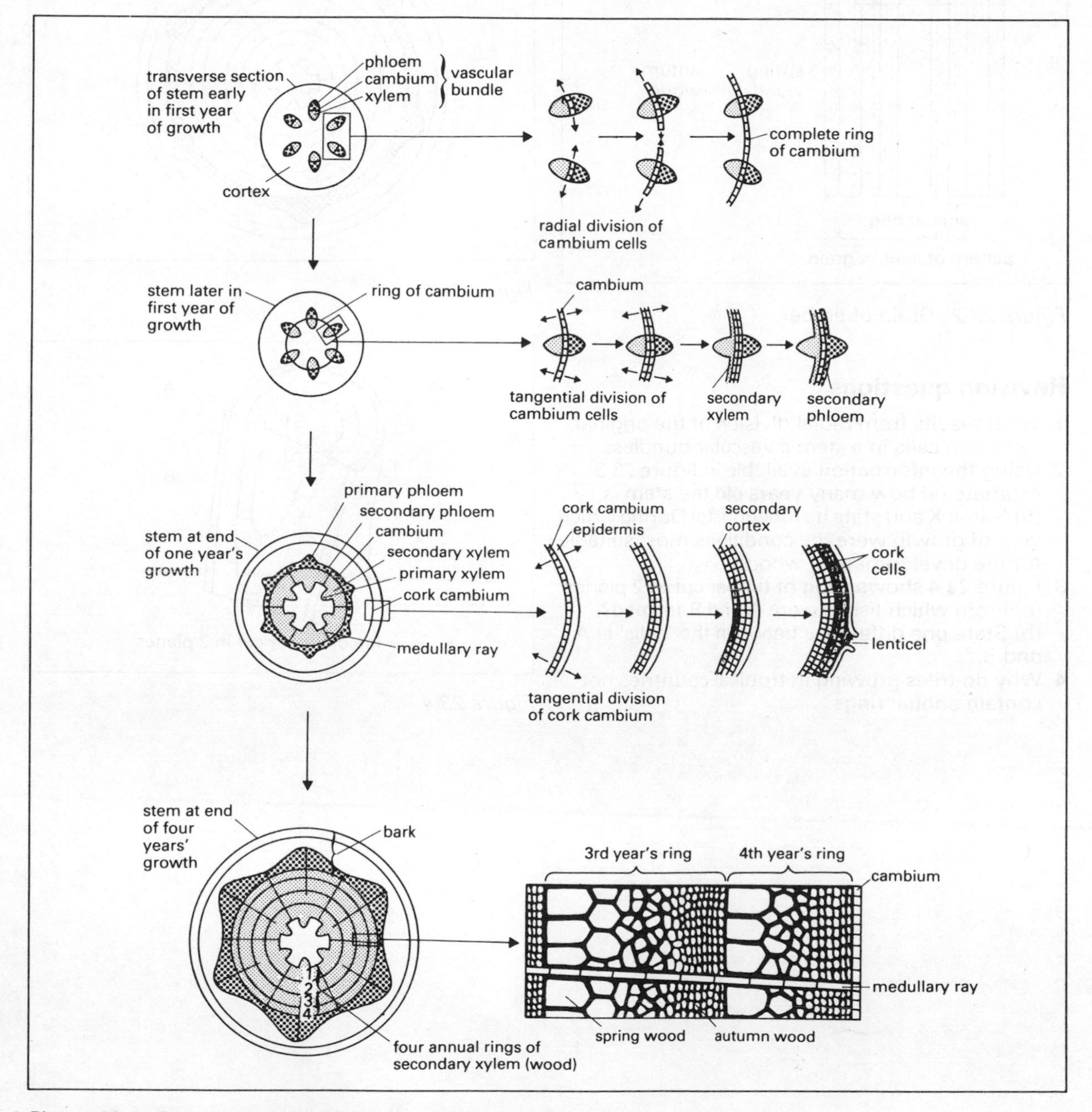

Figure 23.1 Secondary thickening in a woody stem

nally through the vascular tissue, the rings show up as patterns called the **grain** of the wood (figure 23.2). In older trees the innermost xylem vessels (wood) lose their water-conducting capacity and become filled with gums and tannins. This central wood is now called **heartwood** to distinguish it from the outer water-conducting **sapwood**.

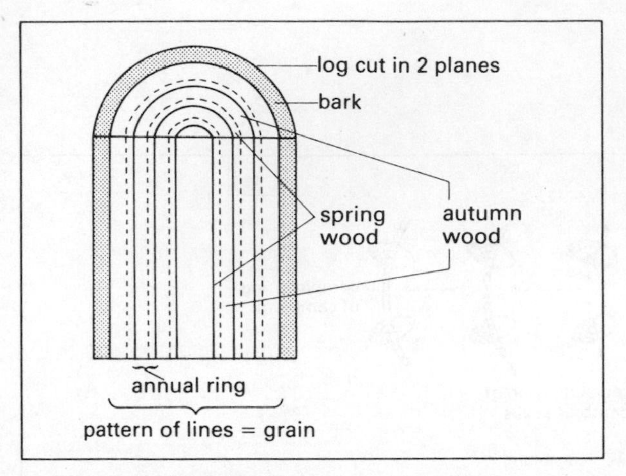

Figure 23.2 Grain of timber

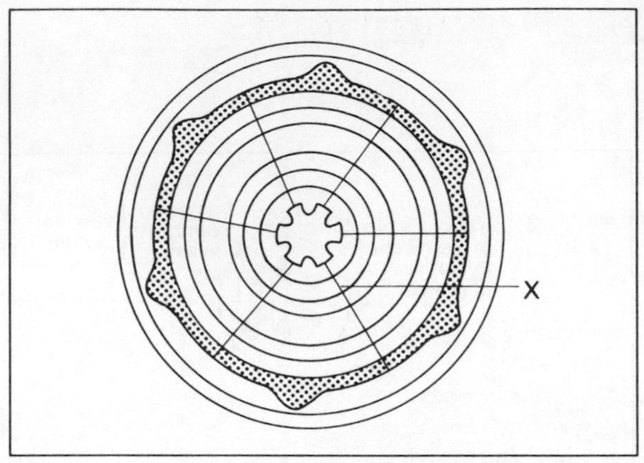

Figure 23.3

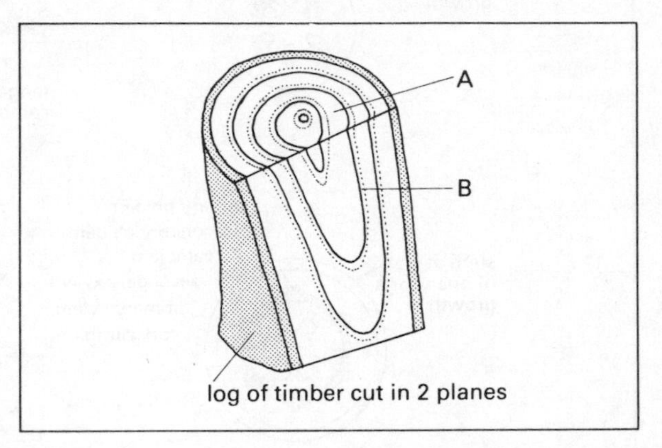

Figure 23.4

Revision questions

1 What results from radial division of the original cambium cells in a stem's vascular bundles?

2 Using the information available in figure 23.3 estimate (a) how many years old the stem is. (b) Name X and state its function. (c) During which year of growth were the conditions most suitable for the development of wood?

3 Figure 23.4 shows a log of timber cut in 2 planes. (a) From which tissue were A and B formed? (b) State one difference between the 'cells' at A and B.

4 Why do trees growing in tropical countries not contain annual rings?

24 Homeorhesis

Homeorhesis is the control of development of an organism by its **genotype** and by the **environment.**

Control of development by genotype

The unicellular alga, *Acetabularia* (figure 24.1), is used to demonstrate this type of control. From experiment 1 (figure 24.2) it can be concluded that a **chemical 'message'** passes from the end containing the nucleus, causing regeneration of the new head. From experiment 2, it can be concluded that a chemical substance produced by the nucleus passes into the stalk and that the stalk receives an amount of this substance sufficient only to regenerate one new head. A second species of this alga (figure 24.3) possesses a head easily distinguishable from the first species. From experiment 3 (figure 24.4) involving both species, it can be concluded that the chemical produced by the nucleus to which the stalk is grafted, determines the type of head regenerated (and renders ineffective any existing chemical).

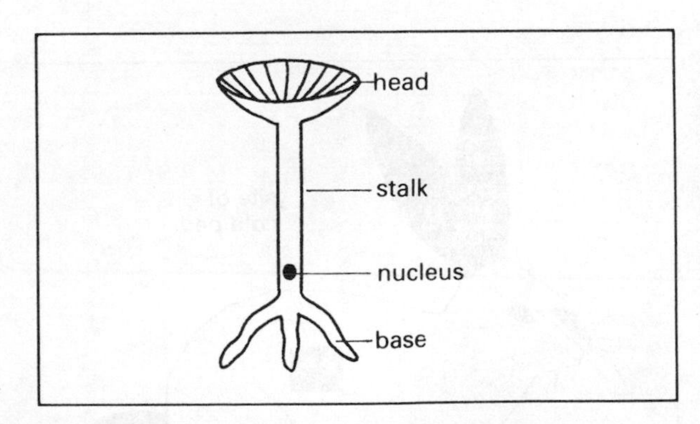

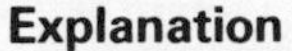
Figure 24.1 Acetabularia mediterranea

Explanation

The DNA of the genes in the nucleus that is present, determines the nature of the **messenger RNA** produced. This mRNA leaves the nucleus and controls the synthesis of the protein required for the regeneration of the new head (see chapter 3). By this means the genotype exerts control over development (and mRNA from any earlier nucleus has no effect).

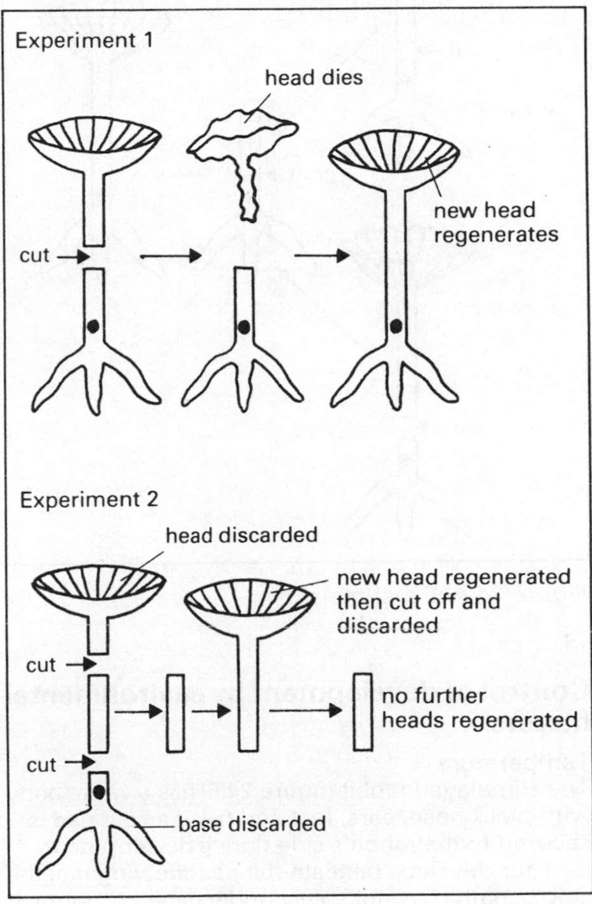

Figure 24.2

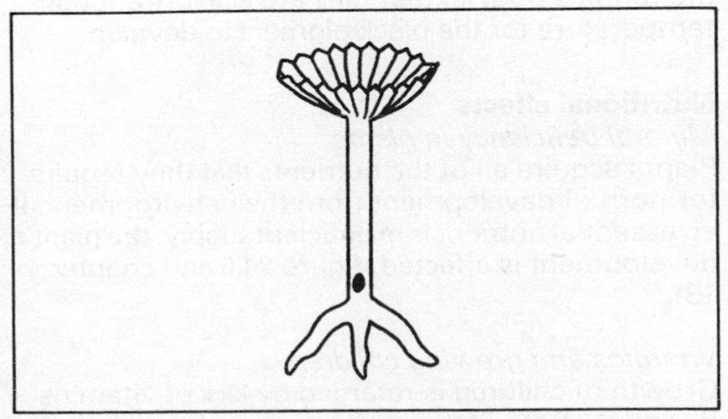

Figure 24.3 Acetabularia crenulata

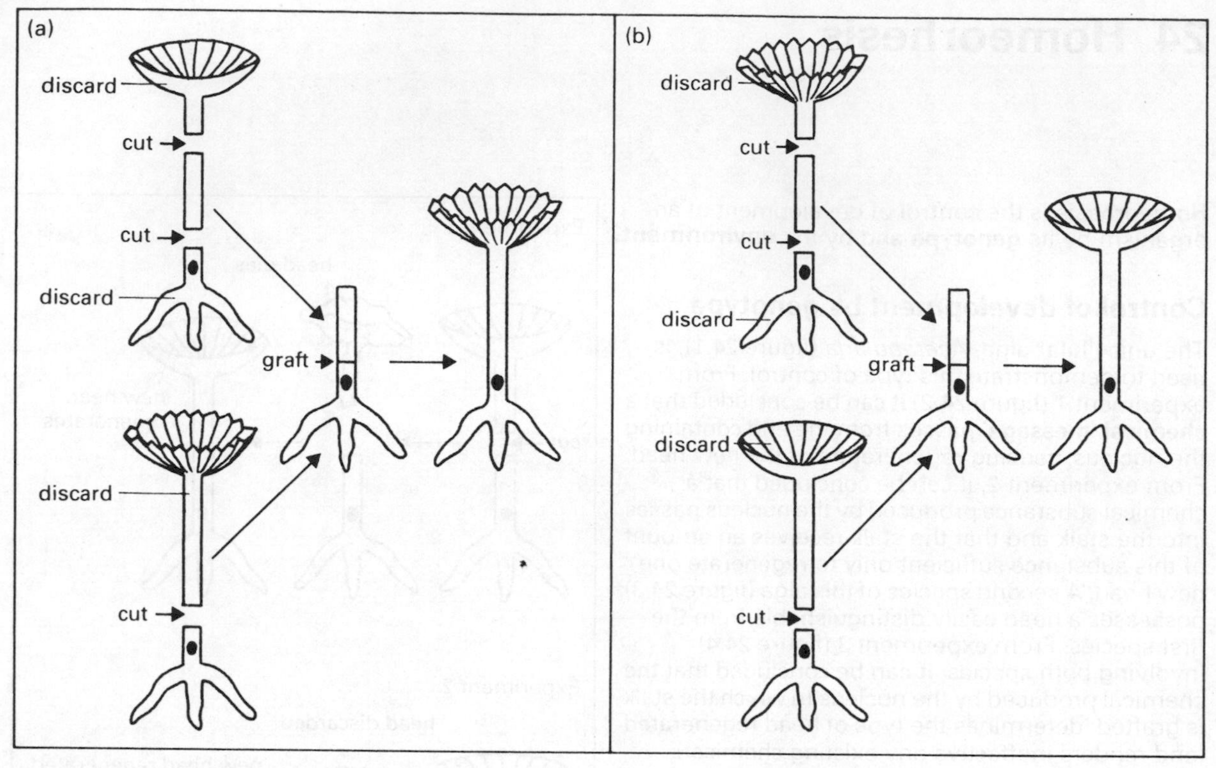

Figure 24.4 Experiment 3

Control of development by environmental factors

Temperature

The Himalayan rabbit (figure 24.5) has a white body with black nose, ears, feet and tail. If a **cold pad** is secured to the rabbit's side during development, dark fur develops beneath the pad showing that the colour pattern is not solely under genetic control. It is now known that heat energy prevents the development of dark pigment (melanin) in this animal and therefore under natural conditions only the rabbit's body extremities are at a low enough temperature for the black pigment to develop.

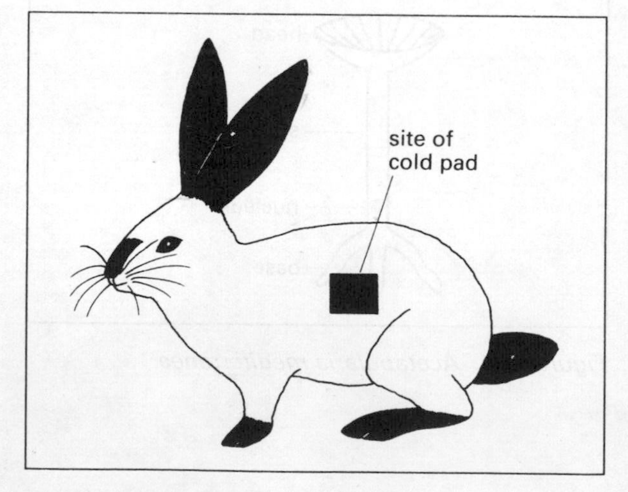

Figure 24.5 Himalayan rabbit

Nutritional effects

Mineral deficiency in plants

Plants acquire all of the nutrients that they require for normal development from their environment. If an essential nutrient is in deficient supply, the plant's development is affected (figure 24.6 and chapter 13).

Vitamins and growing children

Growth of children is retarded by lack of vitamins B_1 and B_2. Both of these vitamins are found in foods such as yeast and liver. The deficiency disease, **rickets** (formation of soft, abnormal bones), is caused by lack of Vitamin D. This vitamin promotes the absorption from the intestine of **calcium** which is essential for the formation of strong bones. Rich sources of vitamin D are cod liver oil, butter and eggs.

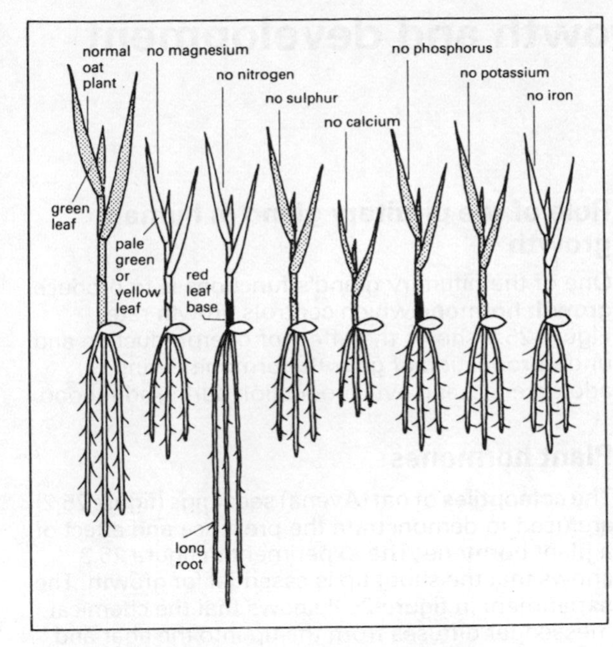

Figure 24.6 Mineral deficiency in plants

Light
When deprived of light, a plant develops abnormally (figure 24.7). The condition that results is called **etiolation**.

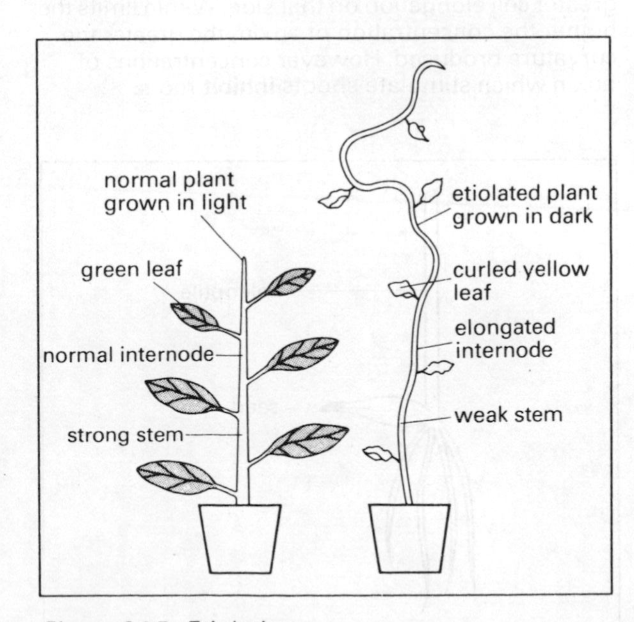

Figure 24.7 Etiolation

Twin studies
The effect of environment on human development can be studied by comparing identical twins who have lived apart in different environments from birth. Since their genotypes are the same, any differences that they show must be due to environmental factors.

Revision questions

1 Complete the following sentence. An organism's phenotype results from the interaction between its . . . and . . ..

2 Draw the resultant structures which would regenerate on bases A and B in the experiment shown in figure 24.8.

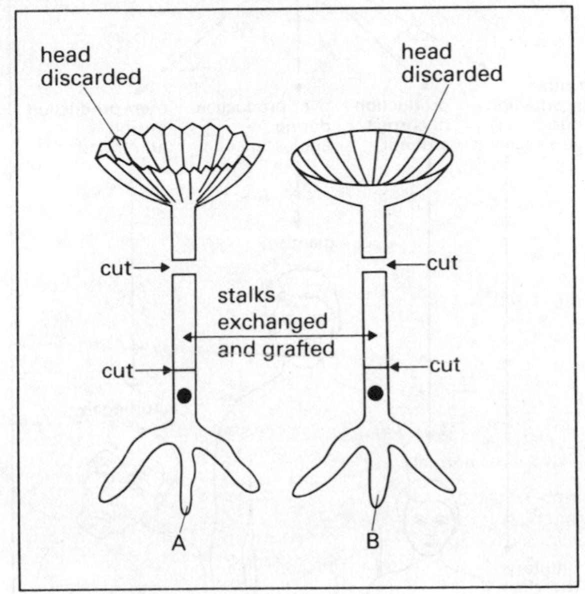

Figure 24.8

3 Name the vitamin required to prevent the deficiency disease rickets.

4 With reference to (a) existing shoots and leaves and (b) new shoots and leaves, briefly describe the effect of transferring a healthy plant into total darkness for several weeks.

5 Identical twin girls aged one year, were orphaned in Japan during World War II. One was adopted by wealthy Americans, the other by poor Japanese relatives. When the girls met again at the age of twenty-one, they were identical in some respects and different in others. (a) List the following under the headings identical and non identical: handspan, bloodgroup, eye colour, body weight. (b) Give one further example of each type of characteristic.

25 Hormonal control of growth and development

A **hormone** is a chemical messenger produced in minute quantity in one part of an organism and transported to another part where it exerts its effect.

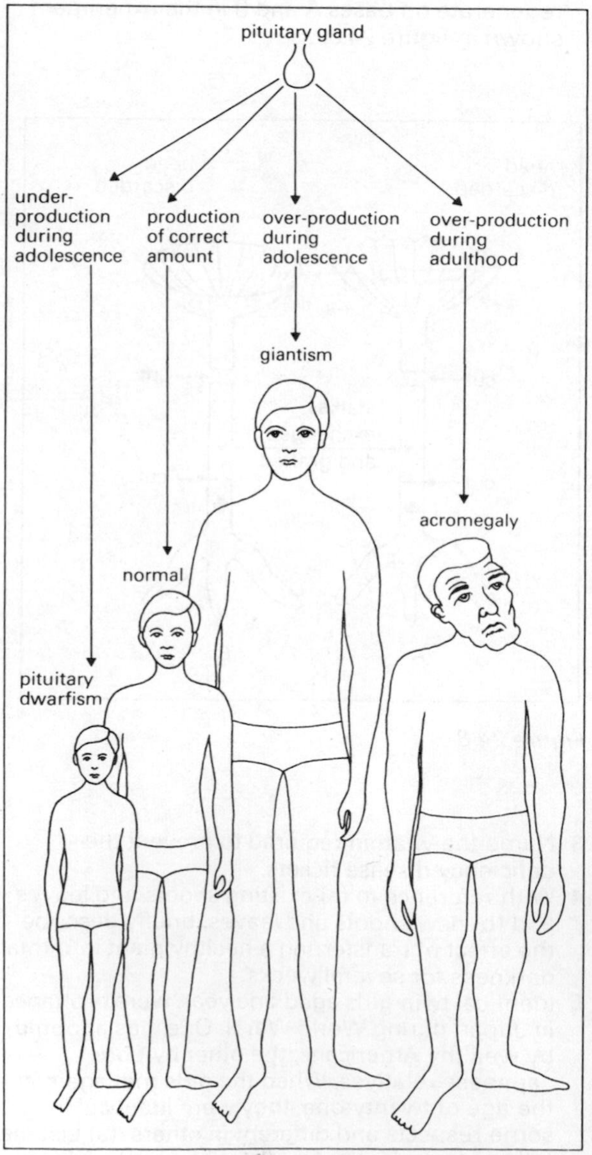

Figure 25.1 Effect of pituitary gland on human growth

Role of the pituitary gland in human growth

One of the pituitary gland's functions is to produce **growth hormone** which controls growth rate. Figure 25.1 shows the effect of overproduction and underproduction of growth hormone during adolescence, and overproduction during adulthood.

Plant hormones

The **coleoptiles** of oat (Avena) seedlings (figure 25.2) are used to demonstrate the presence and effect of a plant hormone. The experiment in figure 25.3 shows that the shoot tip is essential for growth. The experiment in figure 25.4 shows that the chemical messenger diffuses from the tip into the agar and then from the agar into the cut coleoptile where the cells resume **elongation.** The type of hormone which produces this effect is called an **auxin.** The most common auxin is **indole acetic acid (IAA).** In the experiment shown in figure 25.5, the shoot bends because the side below the agar block receives a higher concentration of auxin causing greater cell elongation on that side. Within limits the higher the concentration of auxin, the greater the curvature produced. However concentrations of auxin which stimulate shoots **inhibit** roots.

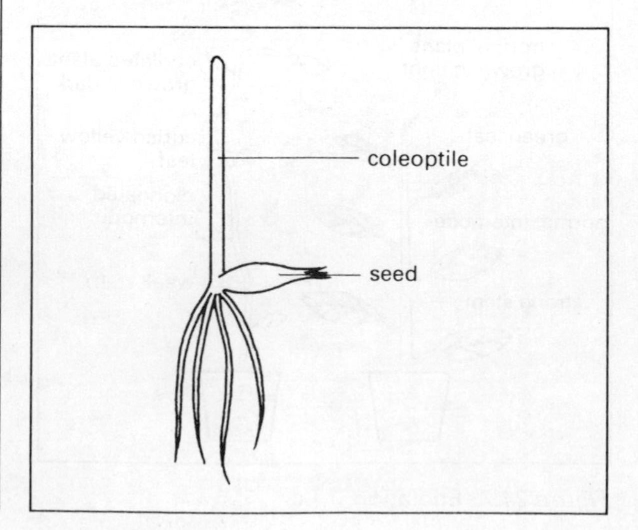

Figure 25.2 Oat seedling

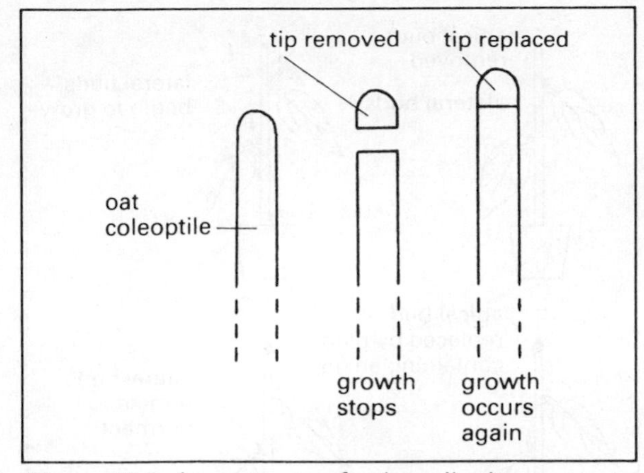

Figure 25.3 Importance of coleoptile tip

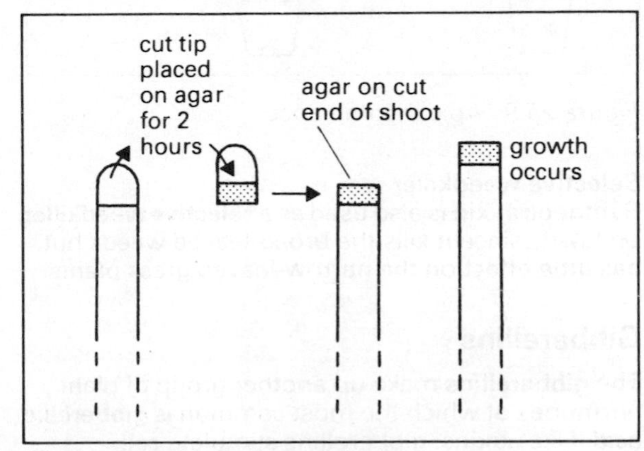

Figure 25.4 Use of agar block

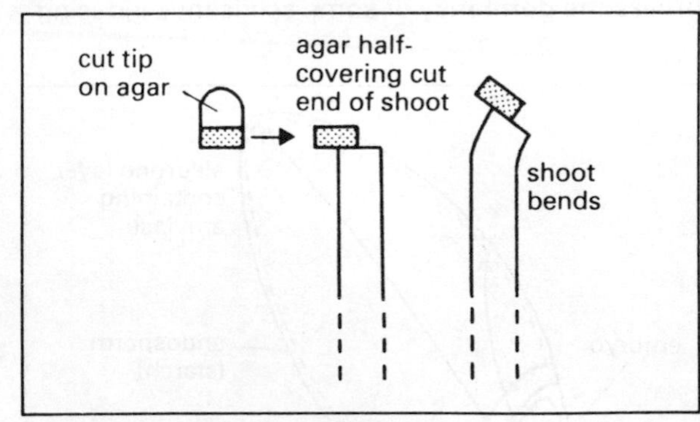

Figure 25.5 Bending effect

Mechanism of phototropism

The experiment in figure 25.6 demonstrates that an oat coleoptile exhibits **positive phototropism** and that the tip is responsible for detecting light. The experiment in figure 25.7 shows that light causes an **unequal** distribution of hormone in the shoot tip. More auxin is present in the non illuminated side

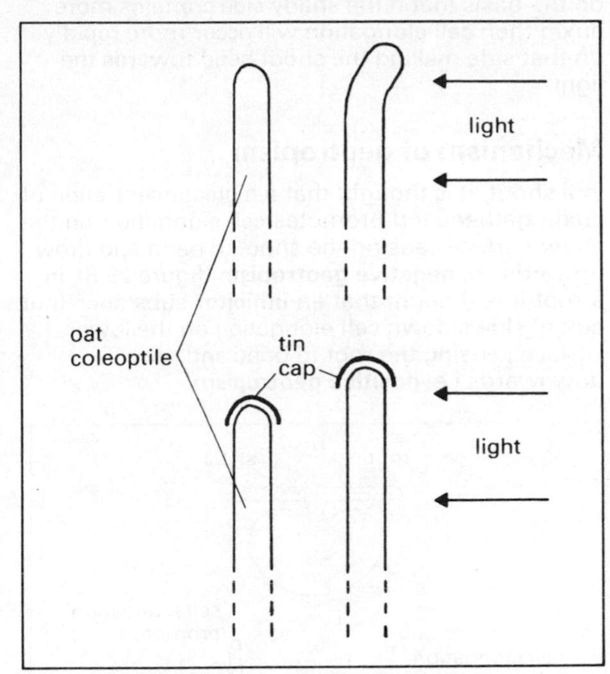

Figure 25.6 Response to light from one side

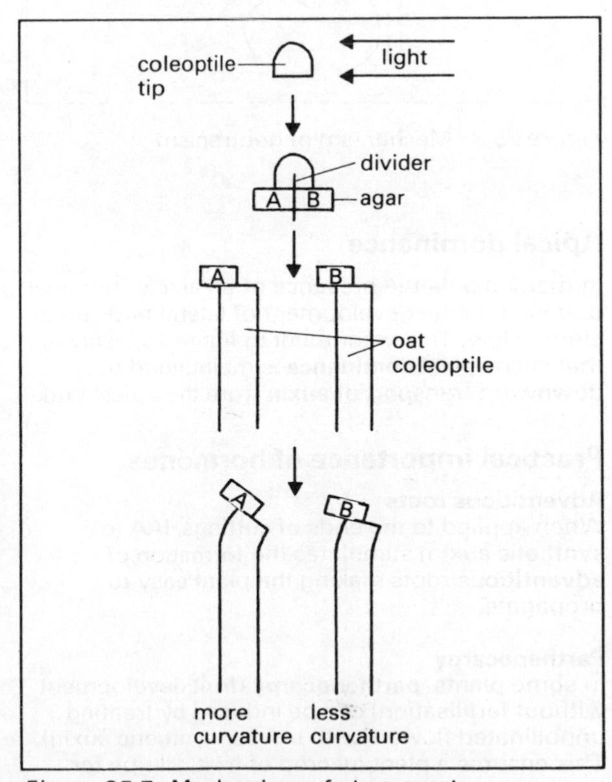

Figure 25.7 Mechanism of phototropism

than in the illuminated side. This information can be used to explain the mechanism of phototropism on the basis that if the shady side contains more auxin then cell elongation will occur more rapidly on that side making the shoot bend towards the light.

Mechanism of geotropism

In a shoot, it is thought that a high concentration of auxin gathers and promotes cell elongation on the lower surface causing the shoot to bend and grow upwards i.e. **negative geotropism** (figure 25.8). In a root it is thought that an **inhibitor** substance (not auxin) slows down cell elongation on the lower surface causing the root to bend and grow downwards i.e. **positive geotropism.**

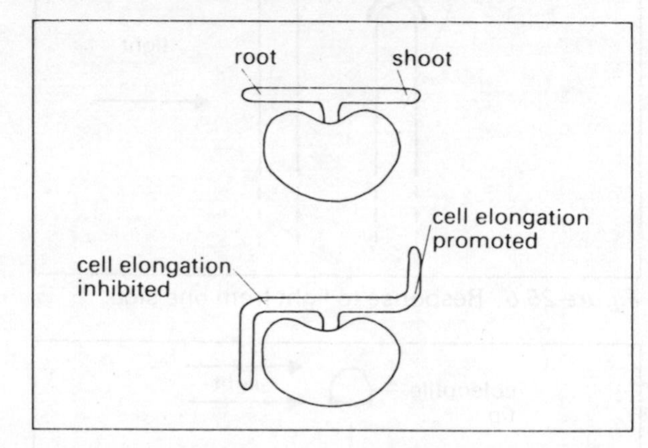

Figure 25.8 Mechanism of geotropism

Apical dominance

In many plants the presence of an apical (terminal) bud inhibits the development of lateral buds on the stem below. The experiment in figure 25.9 shows that such **apical dominance** is maintained by downward transport of **auxin** from the apical bud.

Practical importance of hormones

Adventitious roots

When applied to the ends of cuttings, IAA (or synthetic auxin) stimulates the formation of **adventitious** roots making the plant easy to propagate.

Parthenocarpy

In some plants, **parthenocarpy** (fruit development without fertilisation) can be induced by treating unpollinated flowers with IAA (or synthetic auxin). This ensures a plentiful crop of fruit, all ripe for harvesting at the same time.

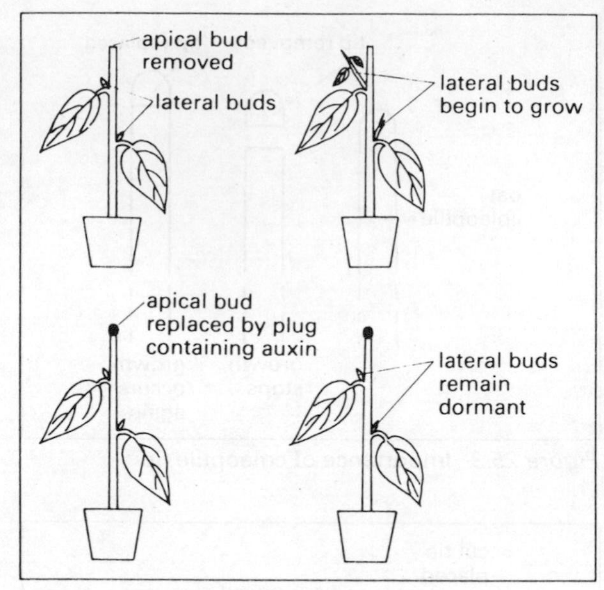

Figure 25.9 Apical dominance

Selective weedkiller

Synthetic auxin is also used as a **selective weedkiller** on lawns since it kills the broad-leaved weeds but has little effect on the narrow-leaved grass plants.

Gibberellins

The **gibberellins** make up another group of plant hormones of which the most common is **gibberellic acid.** Like auxins, gibberellins stimulate **cell elongation** (however they play no part in tropic movements or bending of shoots). Gibberellic acid breaks the **dormancy** of some seeds thus inducing

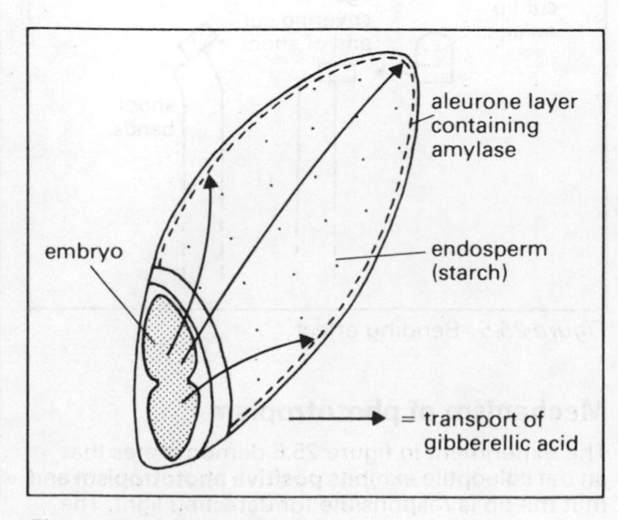

Figure 25.10 Barley seed

them to germinate. In germinating cereal grains (e.g. barley) gibberellic acid is produced by the embryo and passes up into the **aleurone layer** (figure 25.10) where it induces the production of **amylase.** This enzyme converts starchy endosperm to sugar required by the growing embryo.

Revision questions

1 State two differences between acromegaly and pituitary dwarfism.

2 (a) Why are the experiments shown in figures 25.4 and 5 done in darkness?
(b) What controls should be set up for these experiments?

3 Why do the cress seedlings shown in figure 25.11 grow out in a horizontal direction?

4 A soaked barley grain was cut to isolate the embryo region from the endosperm region. The two parts were then placed, cut surface down, on a plate of starch agar as shown in figure 25.12. The procedure was repeated using a plate of starch agar + gibberellic acid. Both plates were left for 24 hours. (a) What colour does starch agar turn on being flooded with iodine solution? (b) Under which one of the four plant parts would this colour change not occur? Explain.

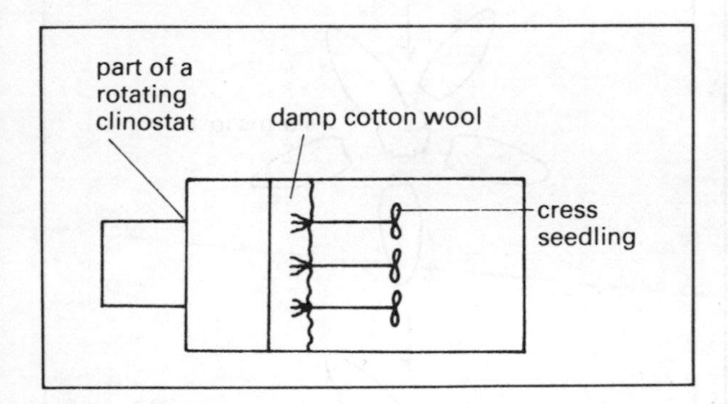

Figure 25.11

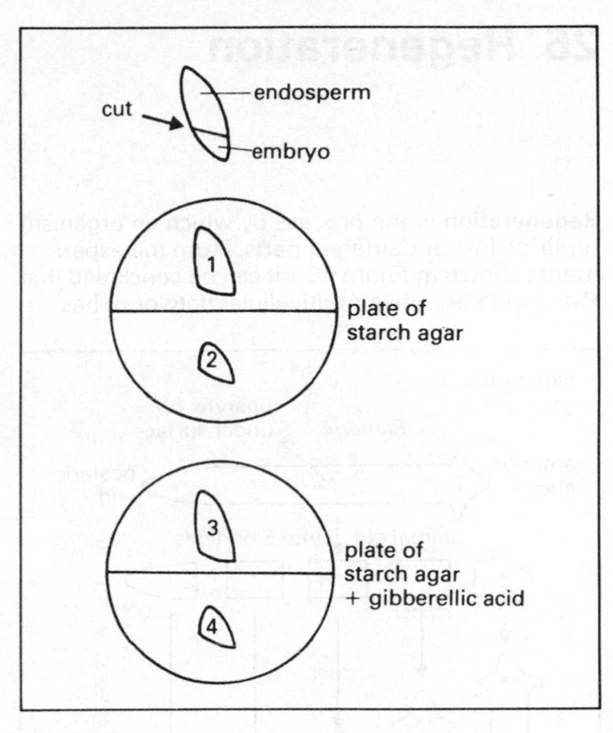

Figure 25.12

26 Regeneration

Regeneration is the process by which an organism replaces lost or damaged parts. From the experiments shown in figure 26.1 it can be concluded that *Planaria* (a primitive multicellular flatworm) has an extensive capacity for regeneration (which decreases from anterior to posterior end). In addition the regenerating pieces retain their **polarity** (i.e. a head grows out from an anterior end, a tail from a posterior end).

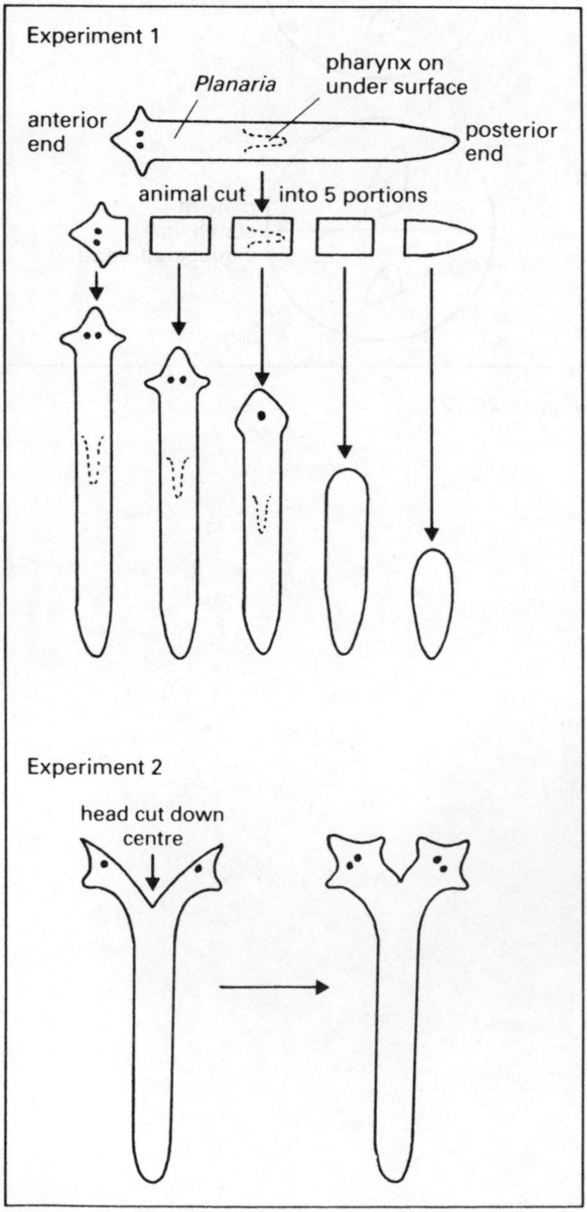

Experiment 1

Planaria
pharynx on under surface
anterior end
posterior end
animal cut into 5 portions

Experiment 2

head cut down centre

Figure 26.1 Regeneration in *Planaria*

Mechanism of regeneration

Distributed throughout *Planaria's* body are numerous **undifferentiated** cells. These cells, which contain a rich supply of mRNA in their

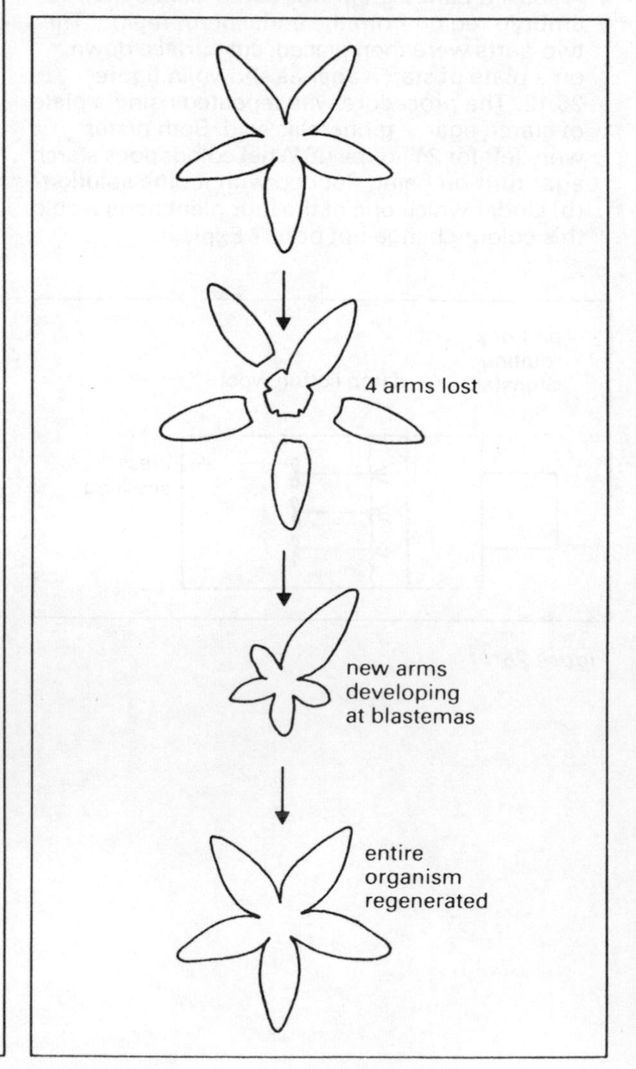

Figure 26.2 Regeneration in starfish

cytoplasm, accumulate at a **wound** surface and form a growing point **(blastema)** which regenerates the missing parts. When a head is regenerated, the first structure formed is the brain, followed by the eyes, then the muscles and lastly the gut.

Regeneration and increasing morphological complexity

A starfish is able to replace a large amount of its body but the **central disc** and at least **one arm** must be present initially for full regeneration to occur (figure 26.2).

Crustaceans and higher animals cannot regenerate the whole organism from just one part. A crab can however regenerate a lost limb (figure 26.3) but when the crab is no longer capable of moulting, regeneration is no longer possible.

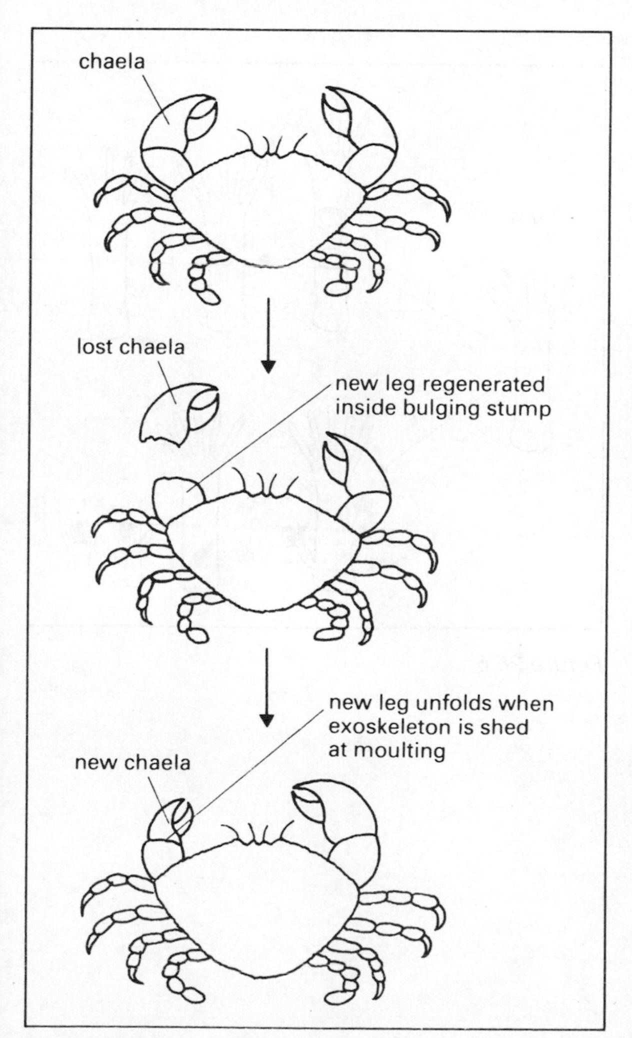

Figure 26.3 Regeneration of a crab's leg

A lizard, when caught by the tail, sheds it in order to escape the predator. A new **tail** is later regenerated.

Birds and mammals cannot replace an entire organ but can regenerate certain damaged or missing parts. Examples include healing of wounds (figure 26.4), mending of broken bones, regeneration of damaged liver and replacement of blood after loss.

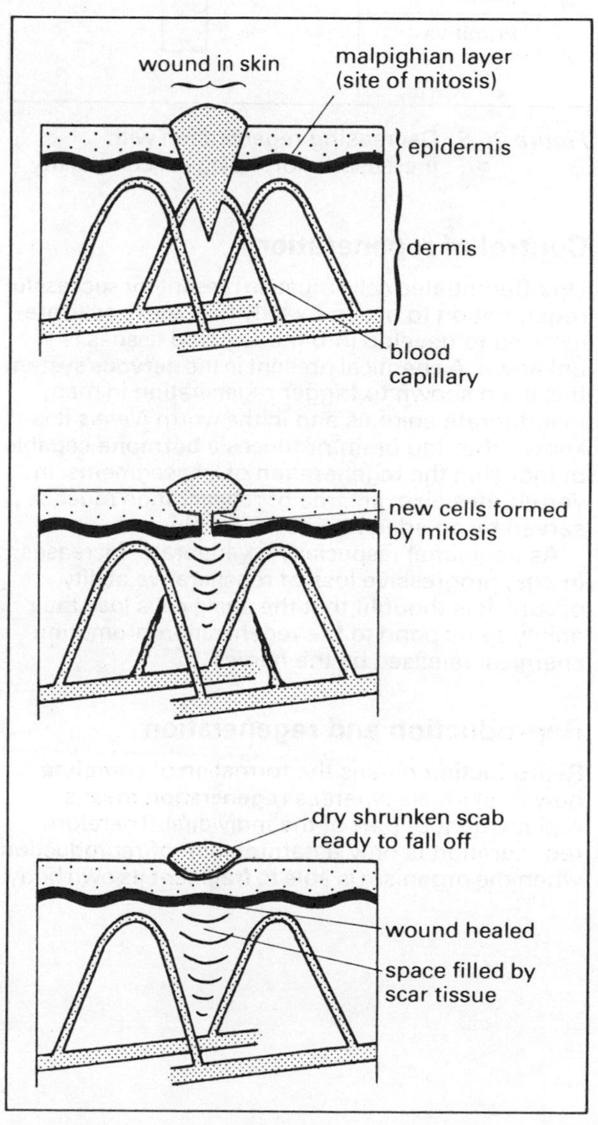

Figure 26.4 Wound healing

Thus increase in morphological complexity is accompanied by decrease in capacity for regeneration as summarised in figure 26.5.

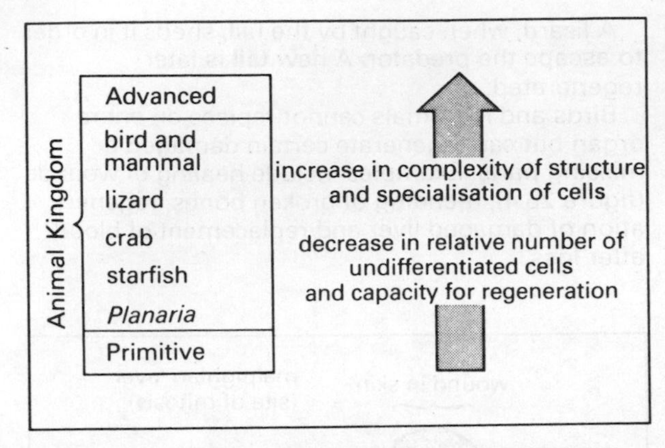

Figure 26.5 Decreasing regeneration with increased morphological complexity

Control of regeneration

Undifferentiated cells must be present for successful regeneration to occur. Exactly how these cells are induced to develop into the required tissues is unknown. A chemical present in the **nervous system** has been shown to trigger regeneration in many invertebrate animals and in the worm *Nereis* it is known that the **brain** produces a **hormone** capable of inducing the regeneration of lost segments. In vertebrates also, an area of regeneration must be served by an adequate nerve supply.

As an animal (especially a vertebrate) increases in age, progressive loss of regenerative ability occurs. It is thought that the aging cells lose their ability to respond to the regeneration-promoting chemical released by the nerves.

Reproduction and regeneration

Reproduction means the formation of **complete** new individuals whereas **regeneration** means replacement of **part** of the individual. Therefore regeneration is only a natural form of reproduction when the organism is able to **fragment** its own body.

For example, the marine worm *Lineus* develops rings of constriction which break the body up into several short fragments. Each fragment then regenerates the whole organism.

Revision questions

1 Name two features common to all animals capable of extensive regeneration.

2 Starfish attack and eat oysters. At one time fishermen used to dredge up starfish, chop them into pieces and throw them back overboard. Was this attempt to protect the oyster beds successful? Explain.

3 With reference to the diagrams in figure 26.6 of *Hydra,* distinguish clearly between reproduction and regeneration.

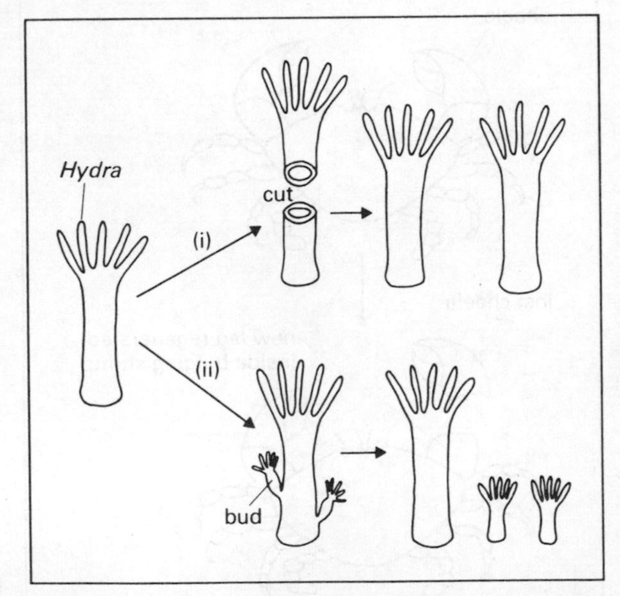

Figure 26.6

27 Photoperiodic effects and cyclic responses

Some plants will suddenly stop producing leaf buds and start producing flower buds in response to a change in the period of illumination **(photoperiod)** to which they are exposed. Such a response to a photoperiod is called **photoperiodism.** The following three distinct groups of plants exist.

Long day (short night) plants
These only flower when the number of hours of light is above the critical level (i.e. the number of hours of darkness is below the critical level) as shown in figure 27.1.

Short day (long night) plants
These only flower when the number of hours of light is below the critical level (i.e. the number of hours of darkness is above the critical level) as shown in figure 27.2.

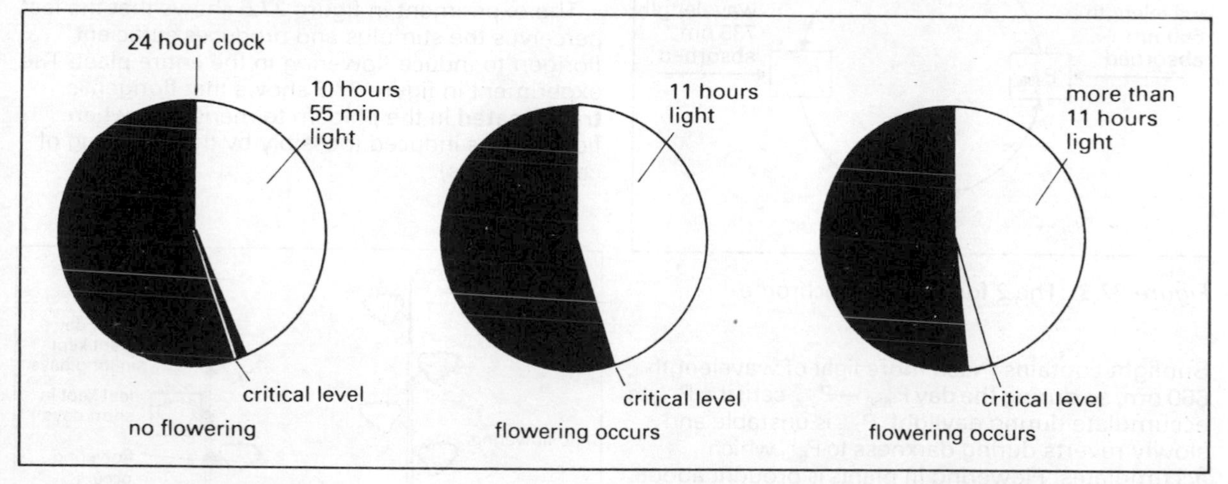

Figure 27. 1 Photoperiodism in long day (short night) plant (e.g. Italian ryegrass—critical duration of light, 11 hours)

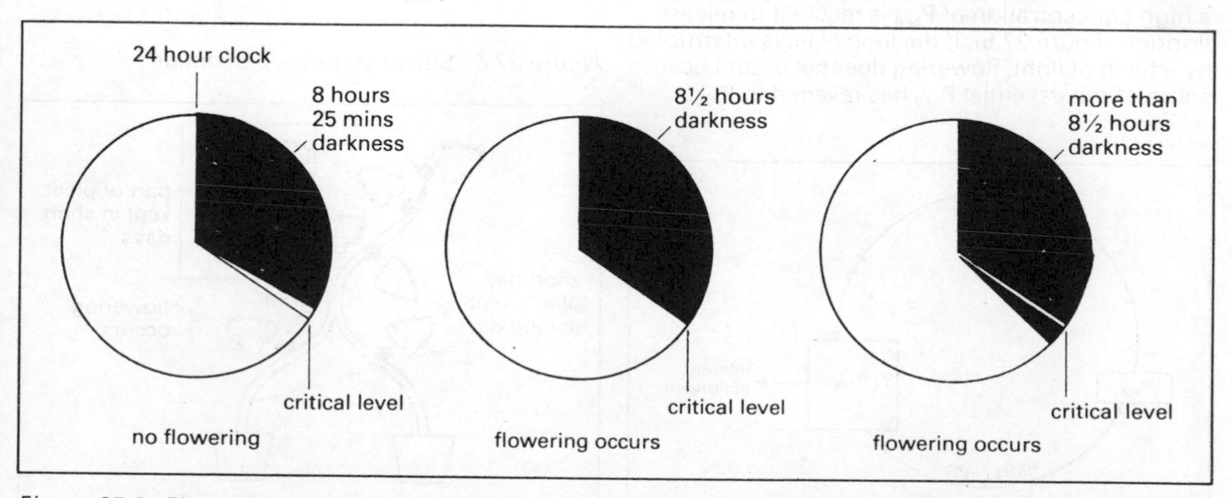

Figure 27.2 Photoperiodism in short day (long night) plant (e.g. cocklebur—critical duration of darkness, 8½ hours)

Day neutral plants
These are plants in which flowering is not dependent upon photoperiod.

Mechanism of the response

Plants contain a chemical called **phytochrome** which exists in two forms. **Phytochrome 660 (P_{660})** absorbs light of wavelength 660 nm and **phytochrome 735 (P_{735})** absorbs light of wavelength 735 nm. Each form changes into the other as shown in figure 27.3.

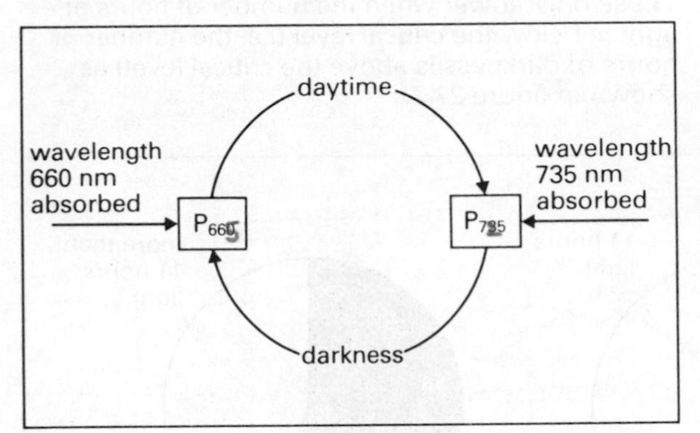

Figure 27.3 The 2 forms of phytochrome

Sunlight contains much more light of wavelength 660 nm, so during the day $P_{660} \rightarrow P_{735}$ causing P_{735} to accumulate during daylight. P_{735} is unstable and slowly reverts during darkness to P_{660} which accumulates. Flowering in plants is brought about by a hormone called **florigen.** In long day plants a high concentration of $\mathbf{P_{735}}$ is required for the release of florigen (figure 27.4). In short day plants a high concentration of $\mathbf{P_{660}}$ is required to release florigen (figure 27.5). If the long night is interrupted by a flash of light, flowering does not occur because some of the essential P_{660} has reverted to P_{735}.

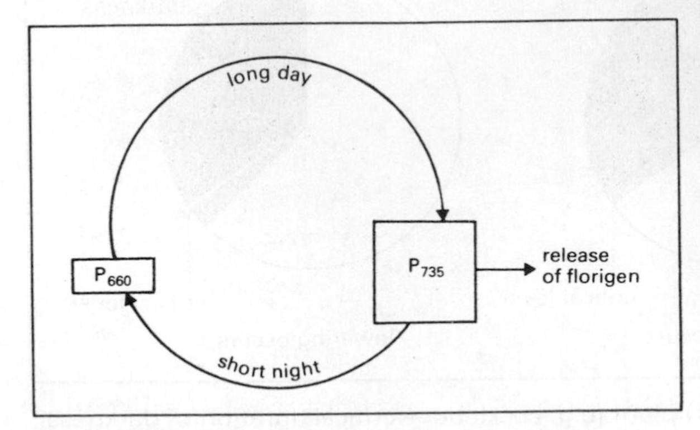

Figure 27.4 P_{735} formation in long day plant

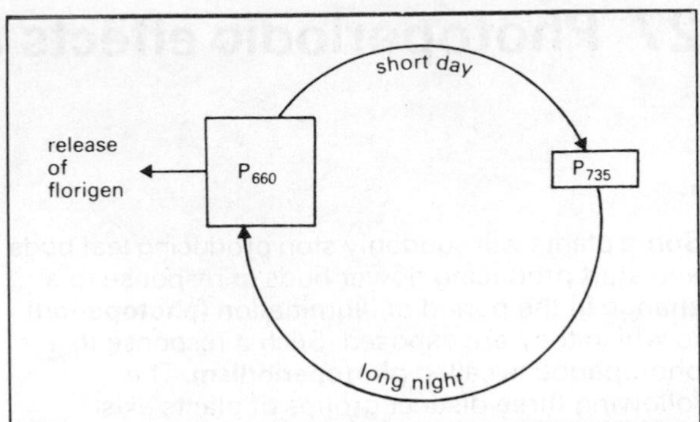

Figure 27.5 P_{660} formation in short day plant

The experiment in figure 27.6 shows that the leaf perceives the stimulus and produces sufficient florigen to induce flowering in the entire plant. The experiment in figure 27.7 shows that florigen is **translocated** in the phloem to meristems where flowering is induced (possibly by the unlocking of certain genes).

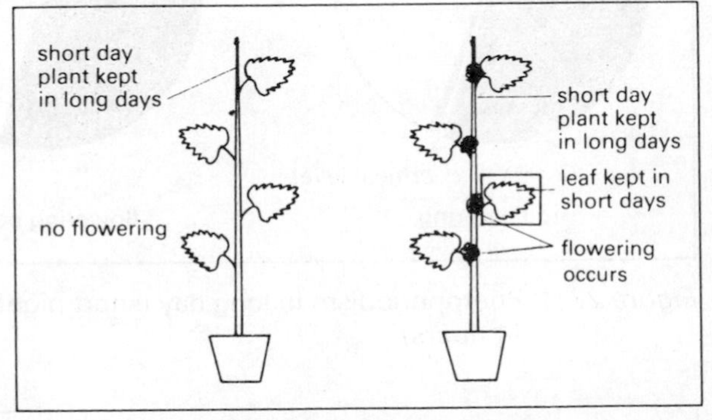

Figure 27.6 Site of stimulus perception

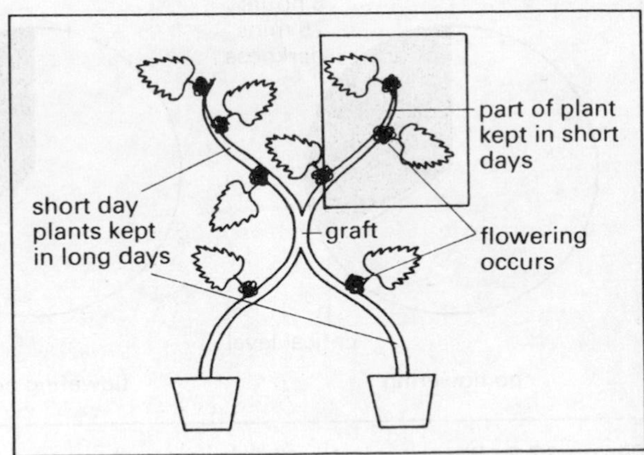

Figure 27.7 Transport of florigen

Latitude and season

The effects of **latitude** and **time of year** on length of photoperiod and therefore on flowering are shown in figures 27.8 and 9.

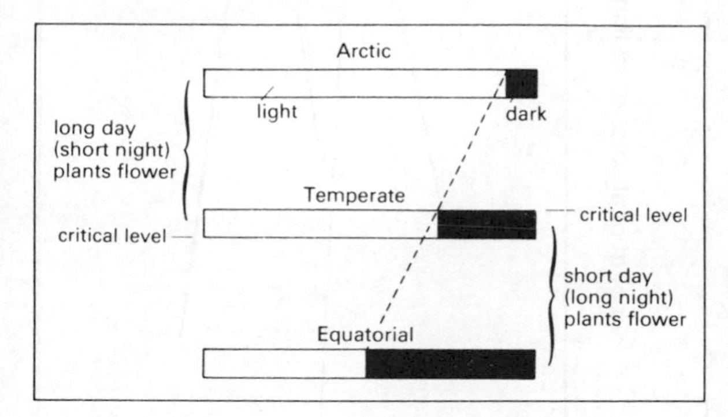

Figure 27.8 Different latitudes (N Hemisphere) in July

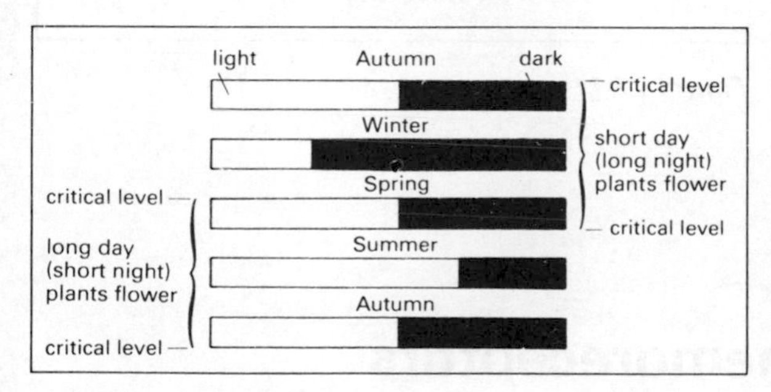

Figure 27.9 Different times of year in temperature latitude

Cyclic responses in animals

Behaviour is described as **cyclic** when it is repeated at definite intervals. Although such behaviour is **endogenous** (under internal control), the time at which it occurs is often influenced by external factors.

Photoperiodic regulation of seasonal cycles

Length of daily **photoperiod** is the only reliable indicator of the time of year to an animal. It is 'measured' by the animal's internal biological 'clock'. Many animals display an **annually recurring** cycle of behaviour by responding once a year to a critical photoperiod. For example, in response to the increasing daylengths that occur in spring, physiological changes (e.g. release of sex hormones) occur within the bodies of many birds, fish (e.g. stickleback) and mammals (e.g. rabbit). Breeding soon follows and offspring are therefore born and reared in favourable environmental conditions. In response to the decreasing day-lengths that occur in autumn, many birds migrate south thus avoiding the rigours of a northern winter and some mammals (e.g. squirrel) begin collecting food in preparation for hibernation.

Circadian cycles

A **circadian cycle** consists of cyclic behaviour which is repeated every 24 hours (e.g. an animal's daily cycle of activity). **Diurnal** animals (e.g. sparrow) show maximum activity during the day and minimum activity at night. **Nocturnal** animals (e.g. owl, cockroach) are the reverse (figure 27.10). Furthermore if a cockroach is kept in constant light, it continues to show the same cycle except that the peak of activity arrives later and later each day since the cycle is now **'free-running'** to its natural length of about 24½ hours. On returning the animal to normal day and night, its activity peak quickly returns to the normal time because the cycle, although endogenous, has become synchronised again to the external factor (light and dark) which is measured by the animal's biological 'clock'. Alteration of the natural timing of a circadian cycle is called **environmental entrainment.**

Some cyclic responses are not synchronised to photoperiod. Grunion fish gather on the beach in California every two weeks at high tide (under lunar control) to spawn and leave fertilised eggs buried in the sand. This two week cycle is in response to the height of the tide/or the phase of the moon. Its advantage is that the eggs develop safely away from predators.

The exact nature of the internal **biological 'clock'** by which an organism measures an environmental stimulus such as photoperiod is still unknown.

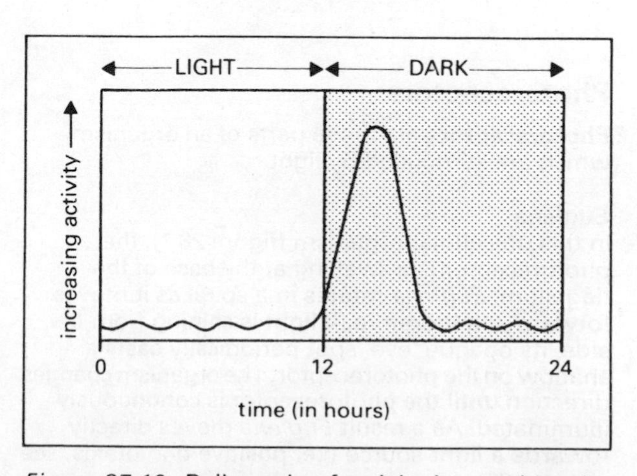

Figure 27.10 Daily cycle of activity in a cockroach

Revision questions

1 (a) Substances A and B shown in the graph (figure 27.11) are found to occur in flowering plants. Identify them.
(b) In which plant organ would they be found?
(c) Is flowering in a long day plant triggered by a large or a small concentration of substance B?

2 Explain why a short day plant subjected to 8½ hours of light in 24, will not flower if a light is flashed on once during the dark period, but will flower normally if light is turned off briefly during the light period.

3 In which latitudes would you not expect to find long day plants flowering?

4 At what season of the year would you expect to find short day plants flowering in Britain?

5 What term is used to mean a diurnal cycle of behaviour by an animal in response to an environmental stimulus?

6 Under natural conditions in northern latitudes, ferrets do not breed from September to March. Describe a possible effect of exposing ferrets to prolonged photoperiods during mid-winter.

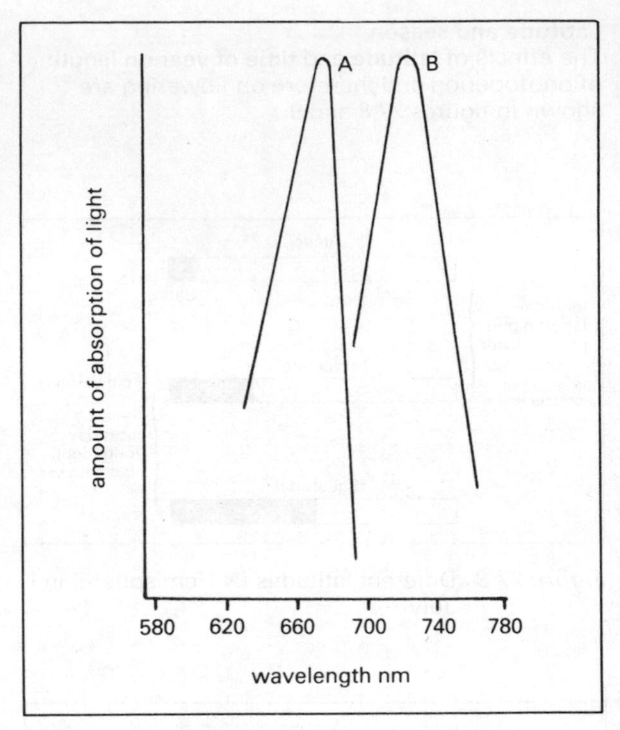

Figure 27.11

28 Photoreceptors and chemoreceptors

Photoreceptors

Photoreceptors are those parts of an organism which are stimulated by light.

Euglena

In this unicellular organism (figure 28.1), the photoreceptor is a swelling at the base of the flagellum. *Euglena* rotates in a spiral as it moves forward and therefore, if light is shining from the side, its opaque 'eye' spot periodically casts a shadow on the photoreceptor. The organism changes direction until the photoreceptor is continuously illuminated. As a result *Euglena* moves directly towards a light source (i.e. positive phototaxis, see chapter 29).

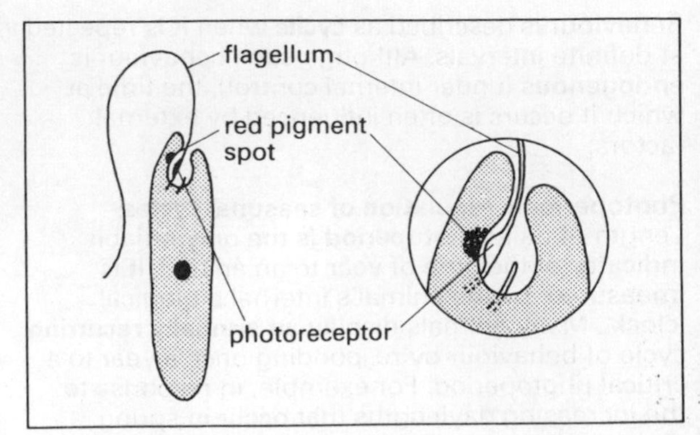

Figure 28.1 Euglena

Earthworm

Photoreceptors are normally **sensory cells** and may be scattered over the whole body as in the earthworm (figure 28.2). Although unable to form definite images, the earthworm is sensitive to light and seeks dark when illuminated.

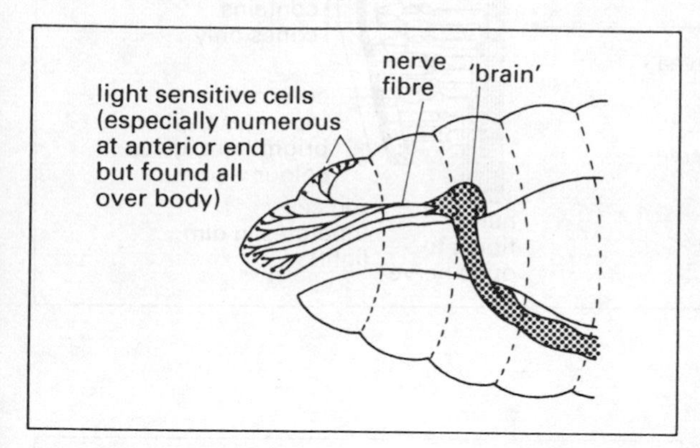

Figure 28.2 Earthworm

More advanced animals

Photoreceptors in more advanced animals are concentrated to form a sense organ, the eye, capable of forming definite images. An insect's **compound** eye (figure 28.3) consists of thousands of separate units, **ommatidia.** Light entering an ommatidium is focused by the lens and crystalline cone and passes down the transparent rod. This is surrounded by light sensitive cells which convert light to electrical impulses transmitted along nerve fibres to the brain. The image seen, depends on the pattern of ommatidia stimulated by light but must be like a picture consisting of mosaic tiles. Much more efficient are the eyes of vertebrates (e.g. the mammalian eye figure 28.4).

Chemoreceptors

Sensory cells which detect the presence of chemicals and as a result, send messages to the animal's brain are called **chemoreceptors.**

Insects

Certain moths have chemoreceptors on their **feathery antennae** (figure 28.5). Chemoreceptors which 'taste' solutions of sugar and nectar are found on houseflies' **bristles** and butterflies' **feet.**

Reptile

A reptile has a **'taste-smell'** organ containing chemoreceptor cells on the roof of its mouth. The forked tongue collects and carries to this organ any scent particles found on the sand (from birds' nests, rodents' burrows etc.). Odours reach this unusual sense organ (figure 28.6) via the nasal cavity. Thus reptiles detect their prey partly by taste and partly by smell.

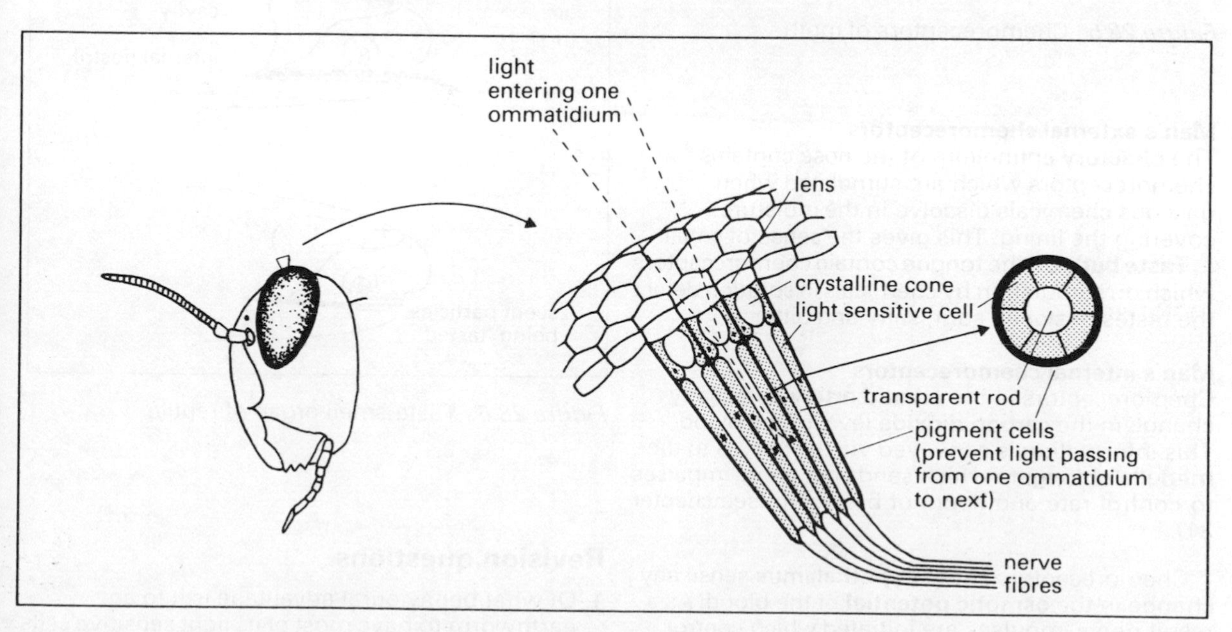

Figure 28.3 Compound eye of insect

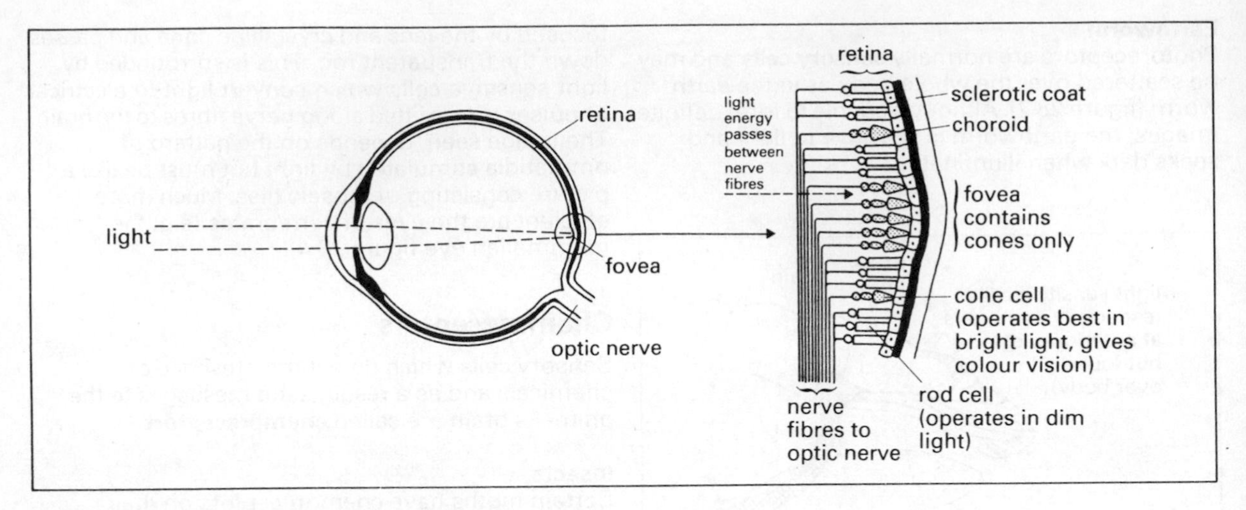

Figure 28.4 Eye of mammal (e.g. man)

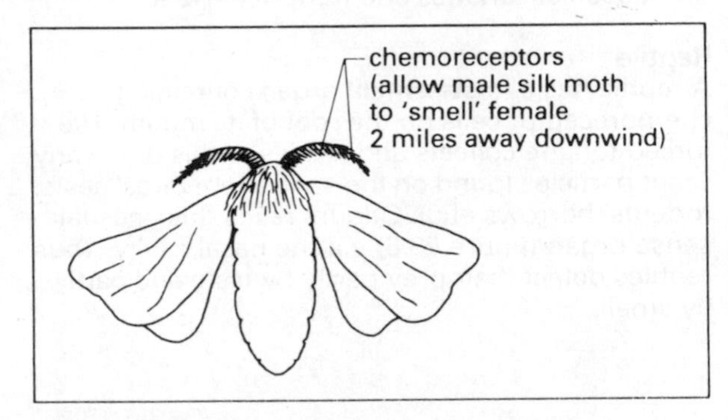

Figure 28.5 Chemoreceptors of moth

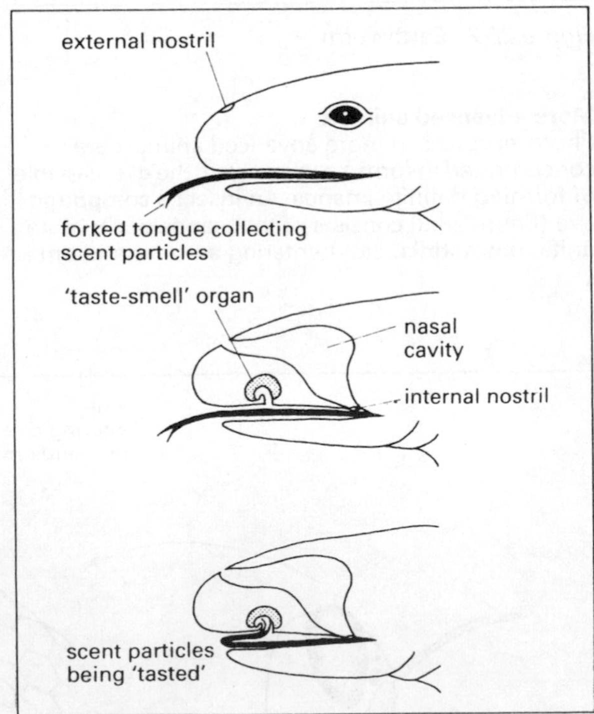

Figure 28.6 Taste-smell organ of reptile

Man's external chemoreceptors
The **olfactory** epithelium of the nose contains chemoreceptors which are stimulated when gaseous chemicals dissolve in the moisture covering the lining. This gives the sense of smell.

Taste buds on the tongue contain chemoreceptors which on stimulation by chemicals in solution detect the tastes of sweet, sour, salty and bitter.

Man's internal chemoreceptors
Chemoreceptors in the **carotid artery** detect any change in the **carbon dioxide** level of the blood. This information is conveyed via the nerves to the **medulla oblongata** which sends out nerve impulses to control rate and depth of breathing (see chapter 31).

Chemoreceptors in the **hypothalamus** sense any change in the **osmotic potential** of the blood. As a result nerve impulses are initiated which control osmoregulatory mechanisms (see chapter 31).

Revision questions

1 Of what behavioural advantage is it to an earthworm to have most of its light sensitive cells at its anterior end?

2 Figure 28.7 shows the image of different shapes falling on the ommatidia of an insect's eye. (a) Draw diagrams shading the ommatidia stimulated by each shape. (b) What can be concluded about the ability of an insect to distinguish shape?
3 Which type of retinal cells would you expect to find in greatest number in the eyes of (a) a nocturnal animal and (b) a diurnal animal?
4 If a butterfly steps on to a drop of sugar solution, its rolled-up proboscis immediately stretches out. Explain.
5 Where are changes in the osmotic potential of man's blood detected?

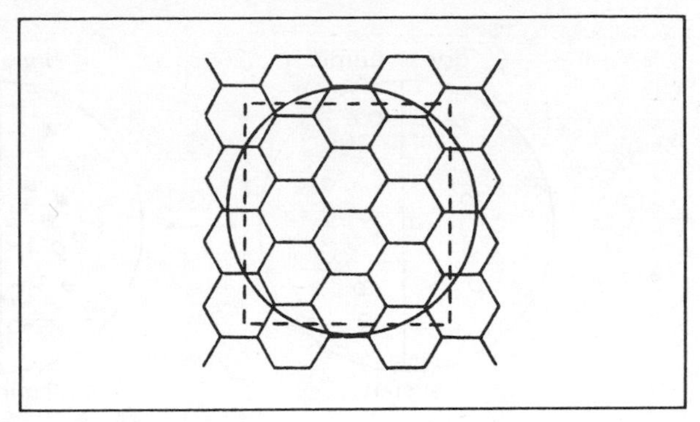

Figure 28.7

29 Unlearned behaviour

Unlearned patterns of behaviour are **inherited, inflexible** responses exhibited by all members of a species in response to particular stimuli.

Kinesis

A **non directional** response to a stimulus is called a **kinesis.** It is demonstrated by woodlice responding to different humidities in a choice chamber (figure 29.1). Figure 29.2 shows the outcome of such an experiment. The woodlice begin randomly distributed. After two minutes those in the dry side increase their rate of **random movement** and rate of **turning** compared to those in the humid side. After four minutes the rate of turning of those in the dry side gradually decreases and they move in long straight lines whereas those in the humid side slow down and may even stop. Due to these differences in their activity woodlice tend to congregate and remain in the favourable environment and move away from the unfavourable one. Similarly, *Paramecia* slow down and congregate in weak acid but speed up and move away from salt solution.

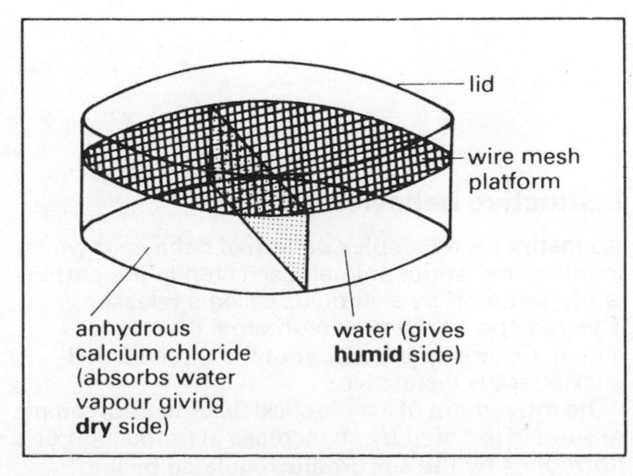

Figure 29.1 Choice chamber

Taxis

A directional response to a stimulus coming from one direction only is called a **taxis.** Movement directly towards the stimulus is called **positive** taxis, movement directly away from it **negative** taxis. Figure 29.3 shows positive phototaxis in *Euglena.* Further examples of taxis are given in table 29.1.

Reflex action

An **automatic unconscious** response by part of the body to a certain stimulus is called a **reflex** action. It is **protective** in nature, for example limb withdrawal, blinking, contraction of pupil in bright light etc.

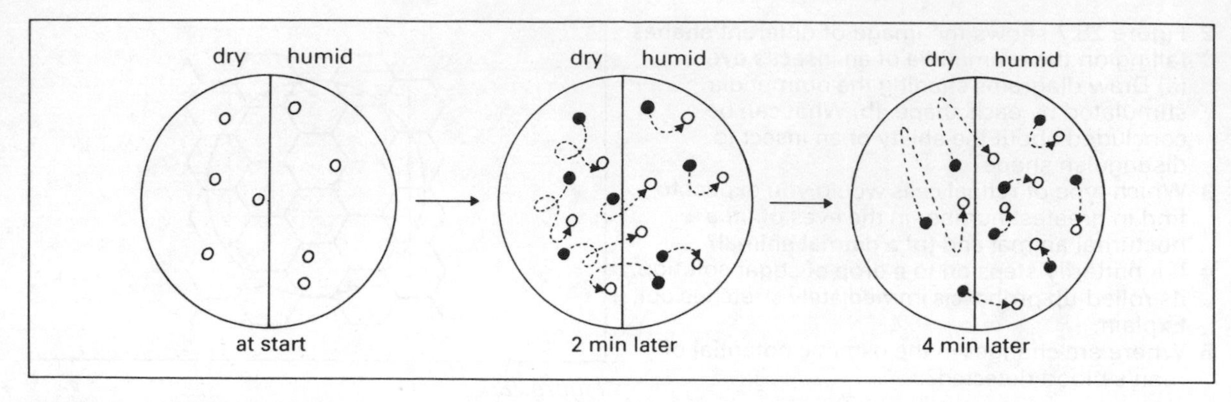

Figure 29.2 Kinesis in woodlice

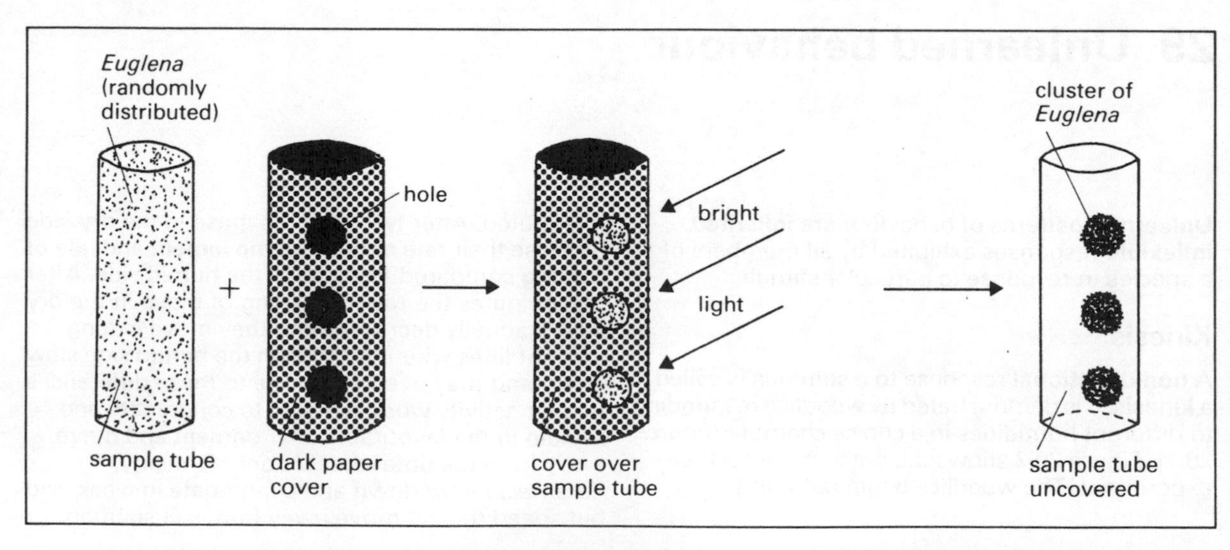

Figure 29.3 Phototaxis in *Euglena*

Animal	Stimulus	Response
maggot	light	negative phototaxis
fruit fly	light	positive phototaxis
planarian worm	chemical (meat)	positive chemotaxis
snail	gravity	negative geotaxis

Table 29.1 Tactic responses

Instinctive behaviour

An **instinct** is a complex pattern of behaviour which involves the entire animal. Each step in the pattern is triggered off by a stimulus called a **releaser**. This may be an external or internal (hormonal) signal. Courtship behaviour of the three-spined stickleback is instinctive.

The movement of a male stickleback to its breeding ground is initiated by an increase in production of hormones by the sex organs regulated by the pituitary gland. This is thought to occur in response

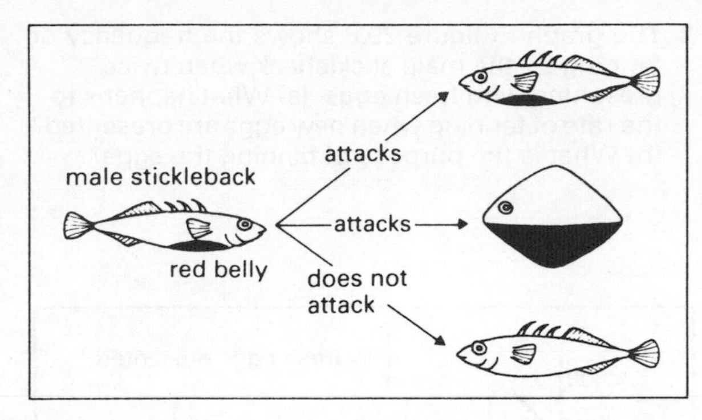

Figure 29.4 Defence of territory by male stickleback

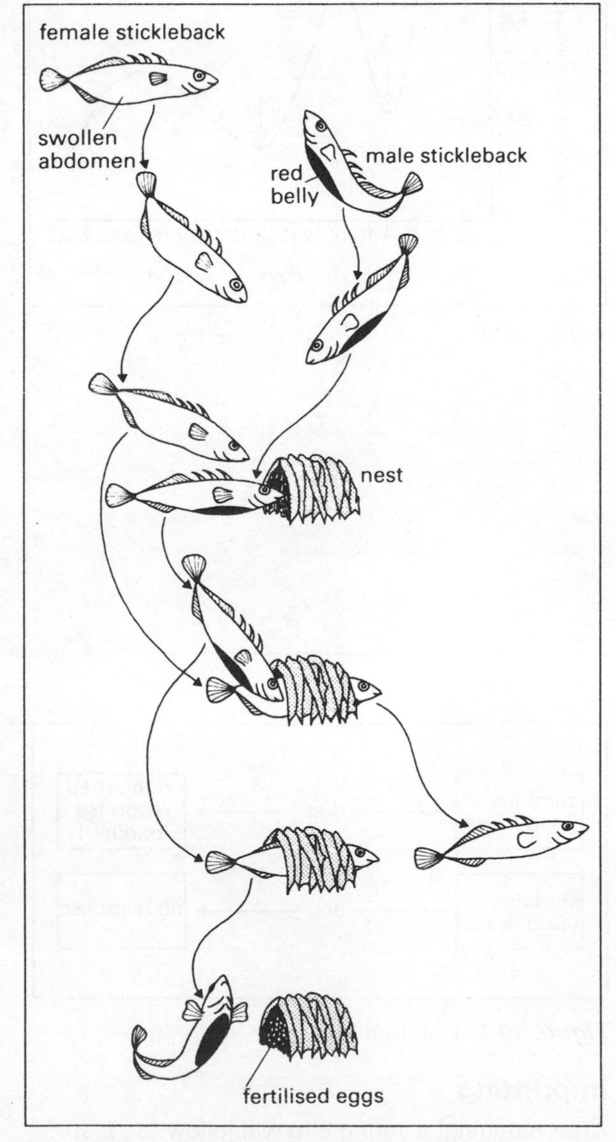

Figure 29.5 Synchronisation in sexual behaviour

to certain external stimuli (e.g. increase in temperature, length of photoperiod). At this time the male's belly is red, indicating that he is in breeding condition. He now chases away rivals, builds a nest of waterweed and then defends the **territory** around it from any intruder with a red belly. He does not attack non red intruders (figure 29.4).

When a female arrives in the territory, the sequence of events shown in figure 29.5 occurs. The female shows her **abdomen** swollen with eggs. On seeing the swollen abdomen, the male performs his **zig-zag** dance (which shows his red belly) and then leads to the nest. Having seen the red belly, the female follows. The male shows the nest and the female enters. The male prods the female's abdomen inducing her to lay the eggs. She now swims away. The presence of the eggs causes the male to enter the nest, fertilise the eggs and then fan them for several days supplying them with oxygen.

This **courtship behaviour** consists of a series of behaviour acts performed alternately by each sex with each act containing the stimulus (releaser) which induces the next act. Such a behavioural pattern synchronises the release of gametes and therefore increases the chance of fertilisation.

Revision questions

1 Figure 29.6 shows the positions of two black Planarian flatworms at 15 second intervals. (a) On which side will the worms tend to congregate? (b) Name this type of unlearned behaviour. (c) Suggest its advantage to the worm.

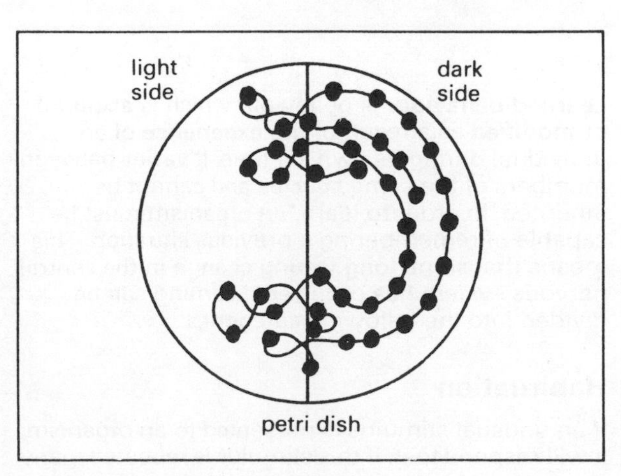

Figure 29.6

2 Figure 29.7 shows models of female sticklebacks. Only 2 and 3 induce courtship behaviour from a male stickleback. (a) Name the stimulus (releaser) to which the male responds. (b) Describe the behaviour act performed by the male. (c) What releaser does this act contain which now affects the female? (d) What behaviour act does she now perform in response?

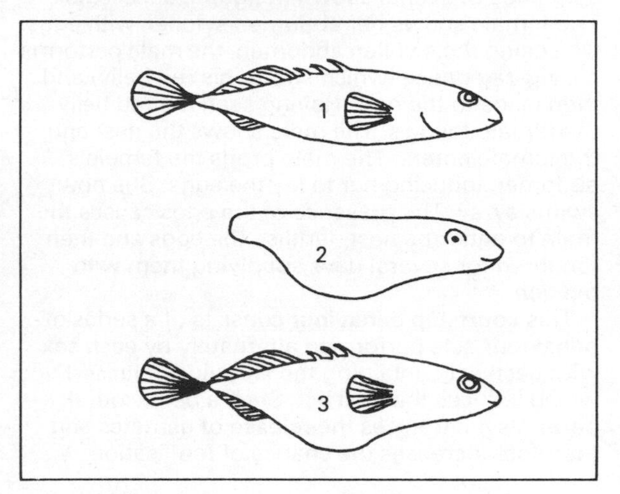

Figure 29.7

3 The graph in figure 29.8 shows the frequency of fanning by the male stickleback when twice presented with fresh eggs. (a) What happens to the rate of fanning when new eggs are presented? (b) What is the purpose of fanning the eggs?

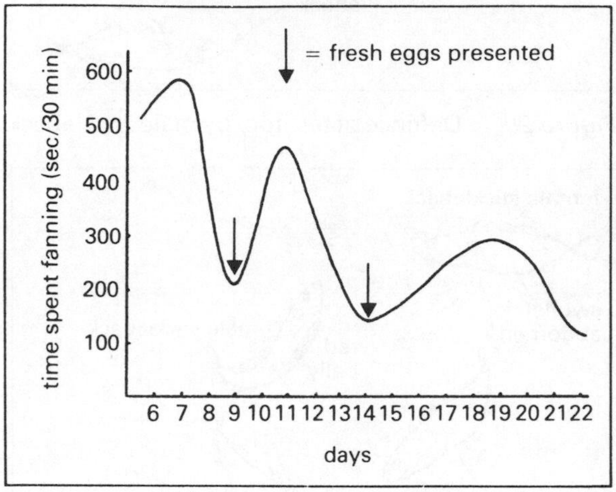

Figure 29.8

30 Learned behaviour

Learned behaviour is behaviour which is **acquired** or **modified** as a result of the **experience** of an individual during its own life time. It **varies** between members of the same species and cannot be inherited. In order to learn, an organism must be capable of remembering a previous situation. This means that some long lasting change in the central nervous system has occurred. Learning can be divided into the following categories.

Habituation

If an unusual stimulus is presented to an organism, it will respond to it. If this stimulus is repeated many times and it proves to be harmless, then the organism will eventually show no response to it (figure 30.1).

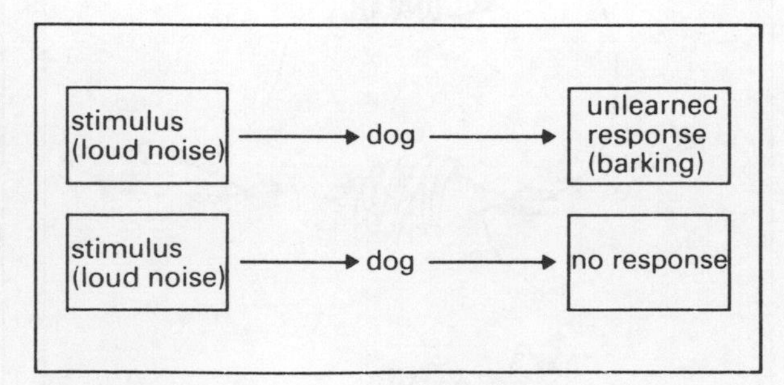

Figure 30.1 Habituation

Imprinting

After hatching, a young bird will follow the first large moving object which it sees. This is normally its mother (figure 30.2). This form of learning which takes place during a critical period of early life is called **imprinting.** The advantages gained by the young bird are that it will be protected by its parent and that it will recognise other members of its own species from which it will later choose a mate.

Figure 30.2 Imprinting

Conditioned reflex

Much of the early work on this type of behaviour was done by the Russian scientist, Pavlov. In the experiment shown in figure 30.3, the dog has learned to respond to a **substitute** stimulus, the bell. The response remains the same. The number of times that the response can be elicited by the **conditioned** stimulus depends on the number of times the food and bell are initially presented together to the dog.

The exact role that this type of conditioning plays in the development of human behaviour is debatable. It is thought, however, that the development of habits and basic skills occurs in a similar fashion. For example, walking, running, talking, cycling, using a knife and fork etc, although all originally learned consciously, later function below the level of consciousness. This leaves the conscious mind free to think about other less routine matters which do require concentration and reasoning.

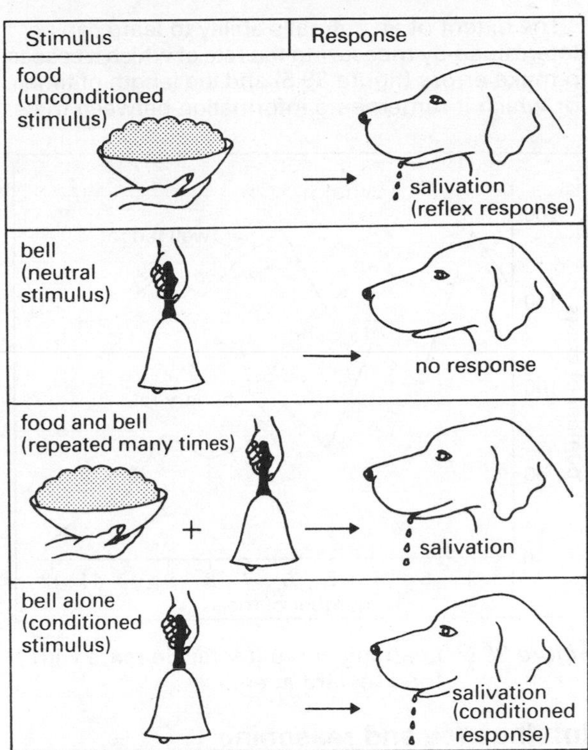

Figure 30.3 Conditioned reflex

Trial and error learning

If a hungry rat is placed in an experimental box (figure 30.4), it will explore the situation and respond in various ways. One of these responses may be to press the lever and as a result be rewarded with food. This is known as **reinforcement.** If the rat is only rewarded when it presses the lever, it soon associates its own behaviour with the **reward.**

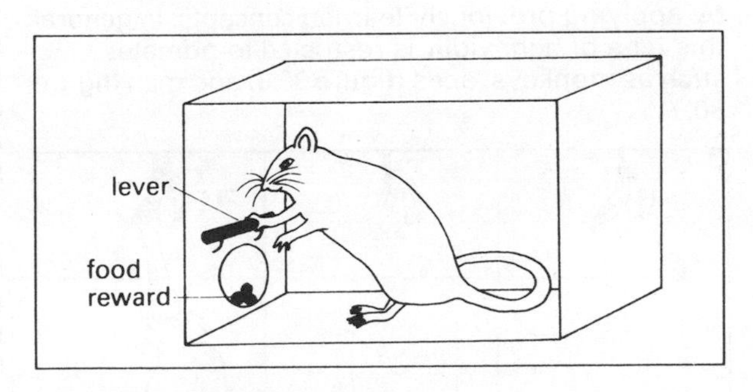

Figure 30.4 Trial and error learning

A hungry rat learns more quickly than one which has been recently fed. The hungry rat is **motivated** i.e. wants the reward. Motivation is the 'drive' which makes the animal want to participate in the learning process. Animals are motivated by many factors (e.g. hunger, thirst, sexual drive, curiosity etc.). **Punishment** (following an incorrect response with an unpleasant sensation e.g. pain) may also act as an effective means of reinforcing the learning process.

The extent of an animal's ability to learn can be determined by measuring the rate of which it ceases to make errors (figure 30.5) and the length of time for which it remembers information between trials.

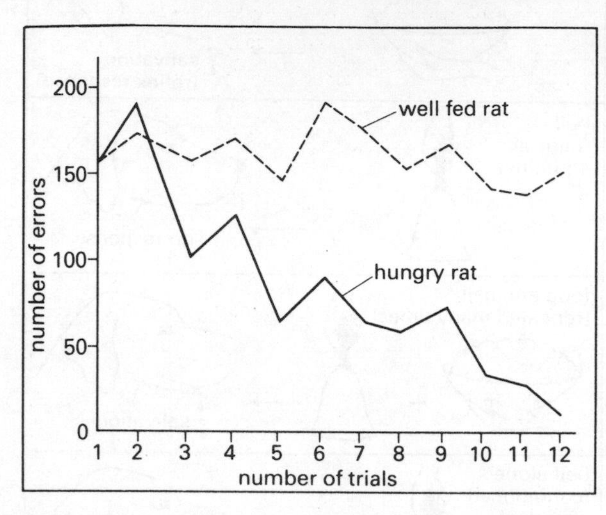

Figure 30.5 Learning curve (for rats in maze with food reward at end)

Intelligence and reasoning

An animal's ability to learn also depends on its **intelligence**. An intelligent animal is capable of **insight learning (reasoning)** and can solve a problem by applying previously learned concepts. In general this type of behaviour is restricted to primates such as monkeys, apes (figure 30.6) and man (figure 30.7).

Figure 30.6 Insight learning (reasoning)

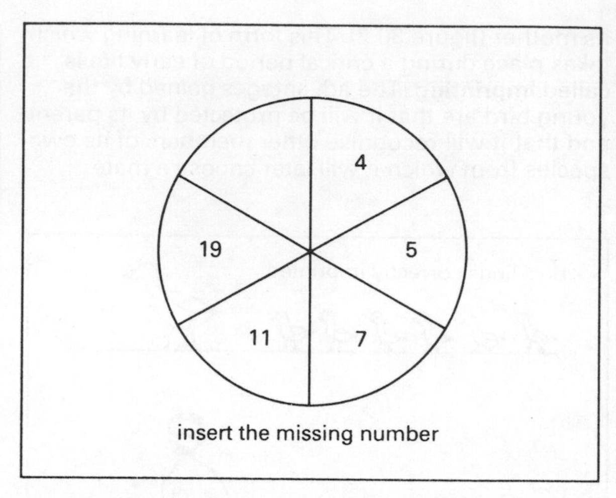

Figure 30.7 Example from an intelligence test

Social behaviour

Many animals live in **social groups** and the interaction of individuals within these groups requires some behavioural adaptation in order to keep the group together. This results in the formation of complex behavioural patterns.

Social signals

Animals use **social signals** to communicate with other members of their own species. A **chemical** signal is given by a **pheromone** (a chemical made by one animal and influencing the behaviour of another). Many vertebrates mark their territory and attract mates by secreting pheromones from their scent glands.

Communication may depend on an **auditory** signal (e.g. bird calls during mating, speech among humans). Visual signals such as colour change (see stickleback chapter 29) and posture of individuals (see revision question 2) may also have an important role to play.

By using such signals, many vertebrate societies establish a **social hierarchy (peck order)** where there is a dominant individual who, due to his rank, has certain rights (e.g. first choice of food, nesting place, available females etc.). Such a peck order decreases the amount of aggression in the group, ensures experienced leadership and promotes the species' chance of survival.

Revision questions

1 The graph in figure 30.8 shows the learning curves of three rats learning to go through a maze.
(a) Identify (i) the hungry rat rewarded with food at the end of each trial, (ii) the well fed rat rewarded with food at the end of each trial,

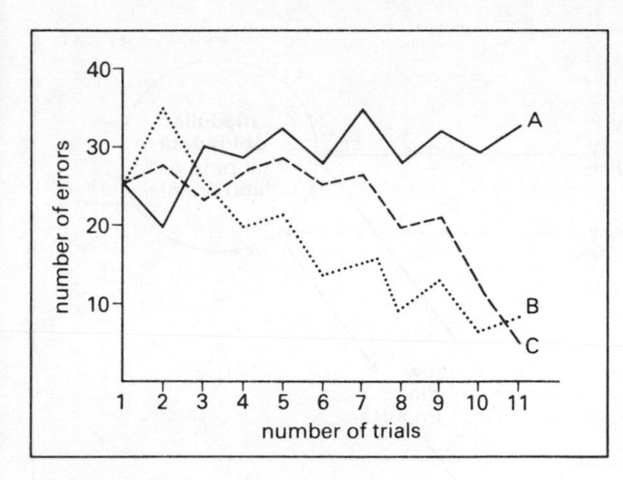

Figure 30.8

(iii) the hungry rat only rewarded with food at the end of trials 6–11.

(b) Which rat was well motivated at the start of the experiment?
(c) Which rat did not learn? Explain why.

2 Instead of fighting, the wolves in figure 30.9 are performing a ritual involving a display.
(a) Describe the state of four of wolf Y's features being used in this display. (b) Describe the state of the same features as displayed by wolf X.
(c) Briefly explain the advantage of this behaviour pattern.

Figure 30.9

31 Homeostasis

Homeostasis is the control and maintenance of a **steady state** in the internal environment of an organism.

Control of blood sugar level

An increase in the concentration of sugar in the blood is detected by the islets of Langerhans in the pancreas which respond by increasing their secretion of the hormone **insulin** into the blood. This increased level of insulin causes the liver to convert the excess blood **sugar** into **glycogen** (and fat) thereby decreasing the blood sugar level. This decrease in turn causes a decrease in insulin secretion which eventually results in an increase in the blood sugar level once more (figure 31.1). This fluctuation around a constant value **(set point)** is an example of homeostasis. This mechanism of regulation is called **negative feedback control.**

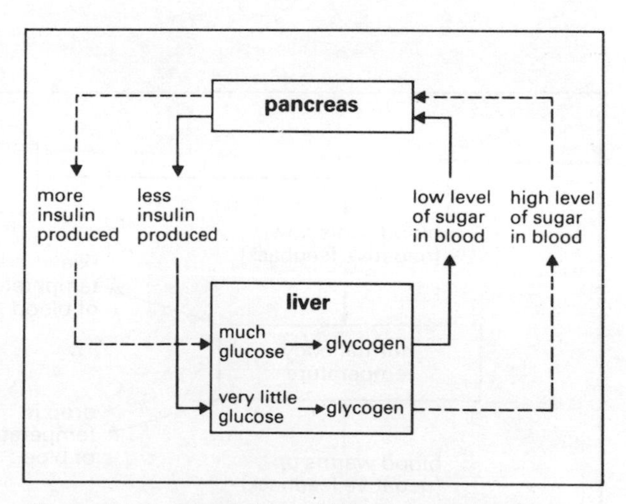

Figure 31.1 Regulation of blood sugar level

Carbon dioxide concentration and regulation of breathing

In some cases part of the homeostatic control is effected via the nervous system, for example carbon dioxide concentration in the blood and regulation of breathing as shown in figure 31.2.

Homeostasis also controls the body's water

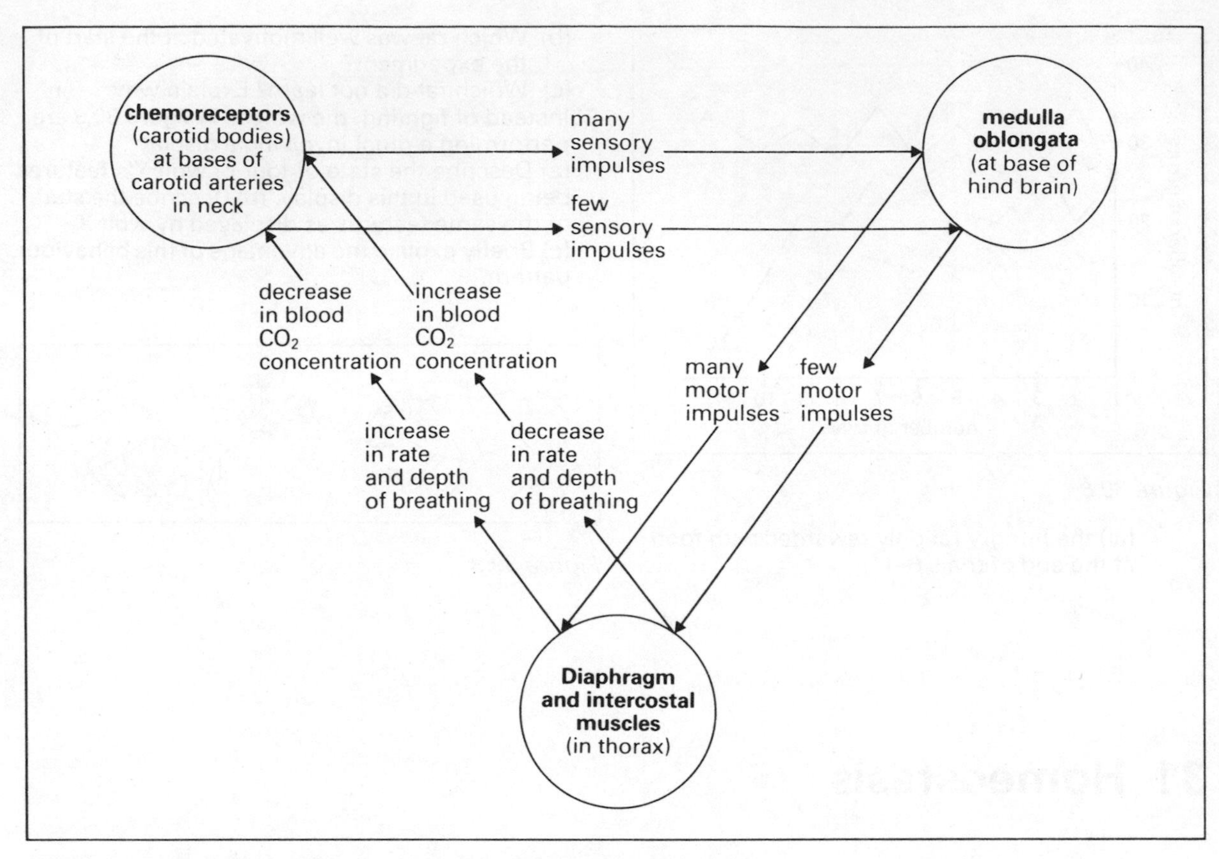

Figure 31.2 Regulation of breathing and CO_2 concentration of blood

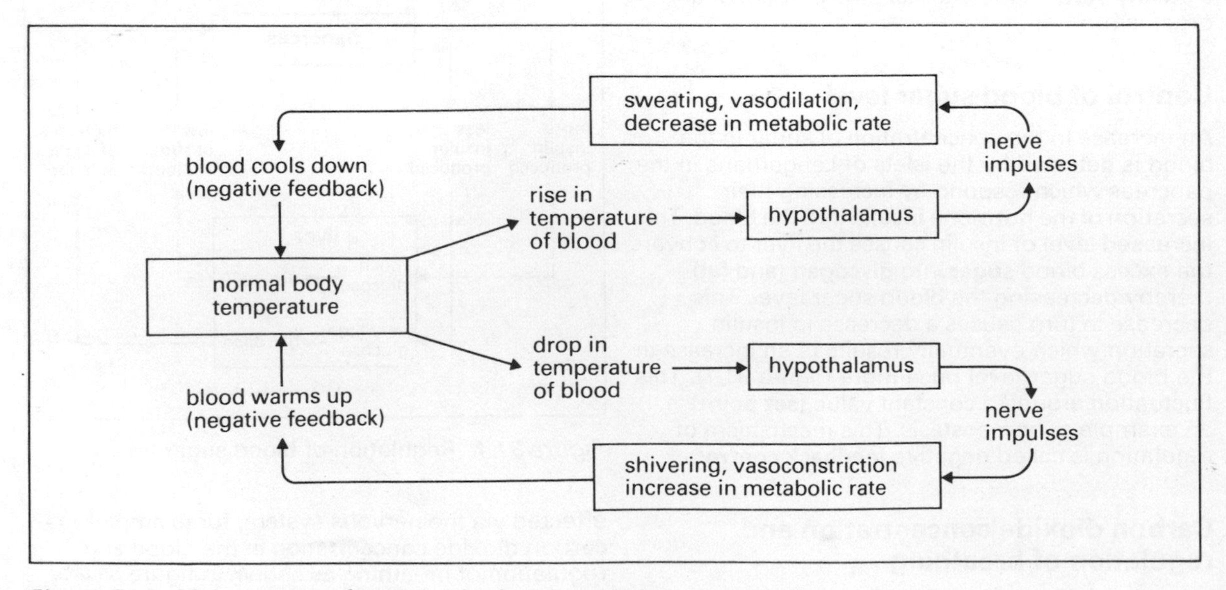

Figure 31.3 Maintenance of constant body temperature

level (see also chapter 11) and maintains the body's constant temperature (figure 31.3). Thus within certain limits, homeostatic control allows the maintenance of a constant internal environment despite wide fluctuations, which are often unfavourable, in the external environment.

Revision questions

1 People suffering from diabetes mellitus excrete much glucose in their urine. What hormone is in low concentration in their blood?

2 % of CO_2 in inspired air

0.04	0.79	2.02	3.05	5.14	6.02

average depth of a breath (cm^3)

673	739	864	1216	1771	2104

average frequency (breaths/minute)

14	14	15	15	19	27

(a) What variable factor was studied in the experiment, the results of which are given above?

(b) Which is affected first, the rate or the depth of breathing?

(c) Which part of the brain responds to an increase in CO_2 concentration in blood?

(d) In response to a message from this part of the brain, two structures react causing an increase in breathing rate. Name them.

3 (a) Which hormone is represented by the letters ADH?

(b) Complete blanks X and Y in figure 31.4 of homeostatic regulation of a mammal's water balance.

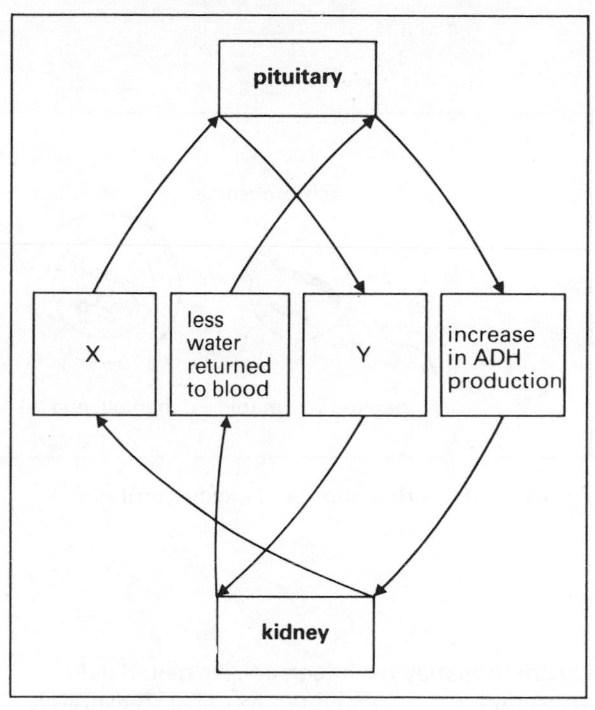

Figure 31.4

32 Meiosis

Haploid and diploid cells

The nucleus of a cell contains a complement of **chromosomes** which varies in number from species to species. A **haploid** cell (e.g. a gamete) has a **single** set of chromosomes (i.e. one of each type) whereas a **diploid** cell (e.g. a zygote) has a **double** set of chromosomes (i.e. two of each type which form **homologous pairs**). A single chromosome complement is represented as n. Therefore a haploid gamete has a ploidy number n and a diploid zygote $2n$. In man $n = 23$ whereas in the example shown in figure 32.1, $n = 2$.

Meiosis

This process results in the production of four haploid (n) gametes from a diploid ($2n$) parental cell. It consists of two consecutive cell divisions and is shown in figure 32.2 for a gamete mother cell containing four chromosomes ($2n = 4$).

During **interphase** each chromosome duplicates forming two identical chromatids. Therefore when the nuclear material condenses during **prophase I,** each chromosome is seen to consist of two chromatids attached at a centromere. During **late prophase I,** homologous chromosomes pair up and

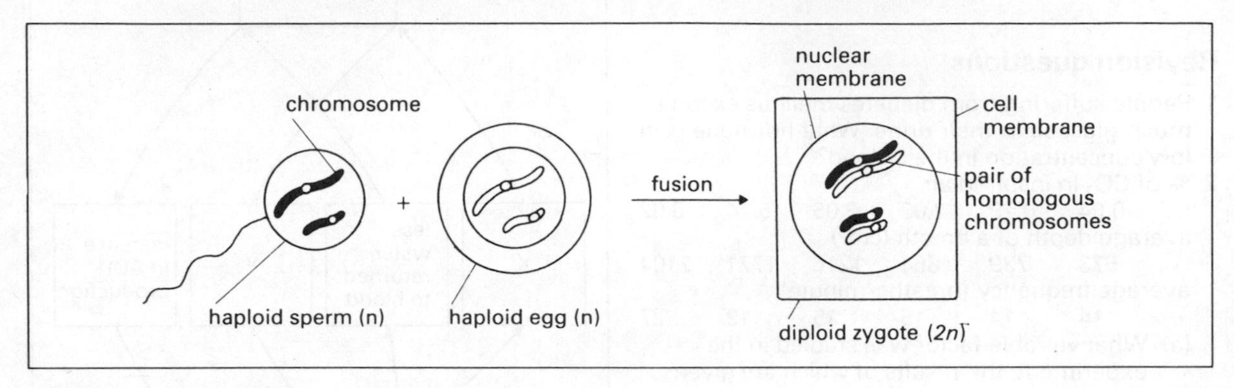

Figure 32.1 Fertilisation and ploidy numbers

chromatids may exchange genetic material if **crossing over** occurs (at points called **chiasmata**). During **metaphase I,** the nuclear membrane disappears, the **spindle** forms from the centrioles and the pairs of homologous chromosomes line up at the **equator**. During **anaphase I,** one member of each homologous (bivalent) pair is drawn to each pole. During **telophase I,** the nuclear membrane reforms and division of cytoplasm occurs. The first meiotic division is called **reduction division** because the two cells formed each contain half of the original number of chromosomes.

The second meiotic division occurs in each of these cells as before except that at **metaphase II,** single chromosomes (each consisting of two chromatids) line up at each equator. Each chromatid, on separation from its partner, becomes a chromosome. The four haploid gametes formed are all different, if crossing over has occurred.

Random assortment of chromosomes

At metaphase I, there is an equal chance that the pairs of homologous chromosomes become arranged as shown in figure 32.3. As a result of this **random assortment** of chromosomes, half of the gamete mother cells produce gametes different to those formed in figure 32.2.

Meiosis occurs wherever gametes are produced e.g. in the **testes** and **ovaries** of animals and in the **anthers** and **ovaries** of flowering plants.

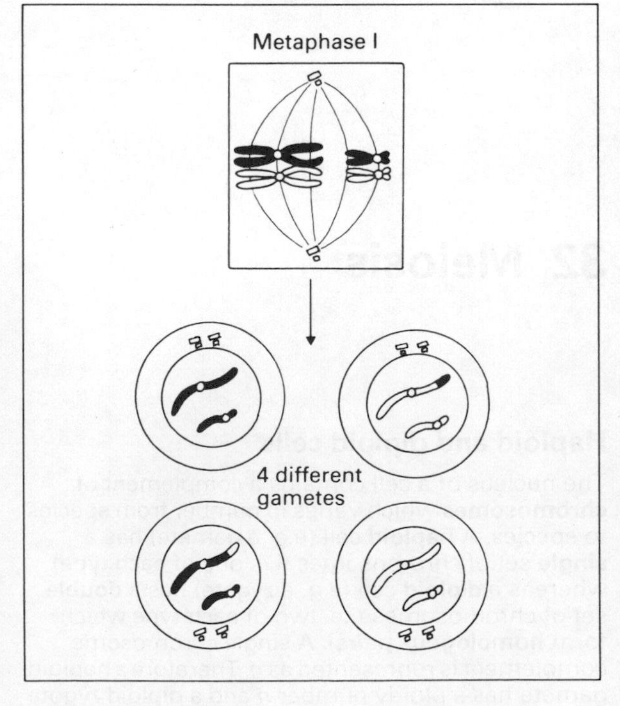

Figure 32.3 Effect of random assortment of chromosomes

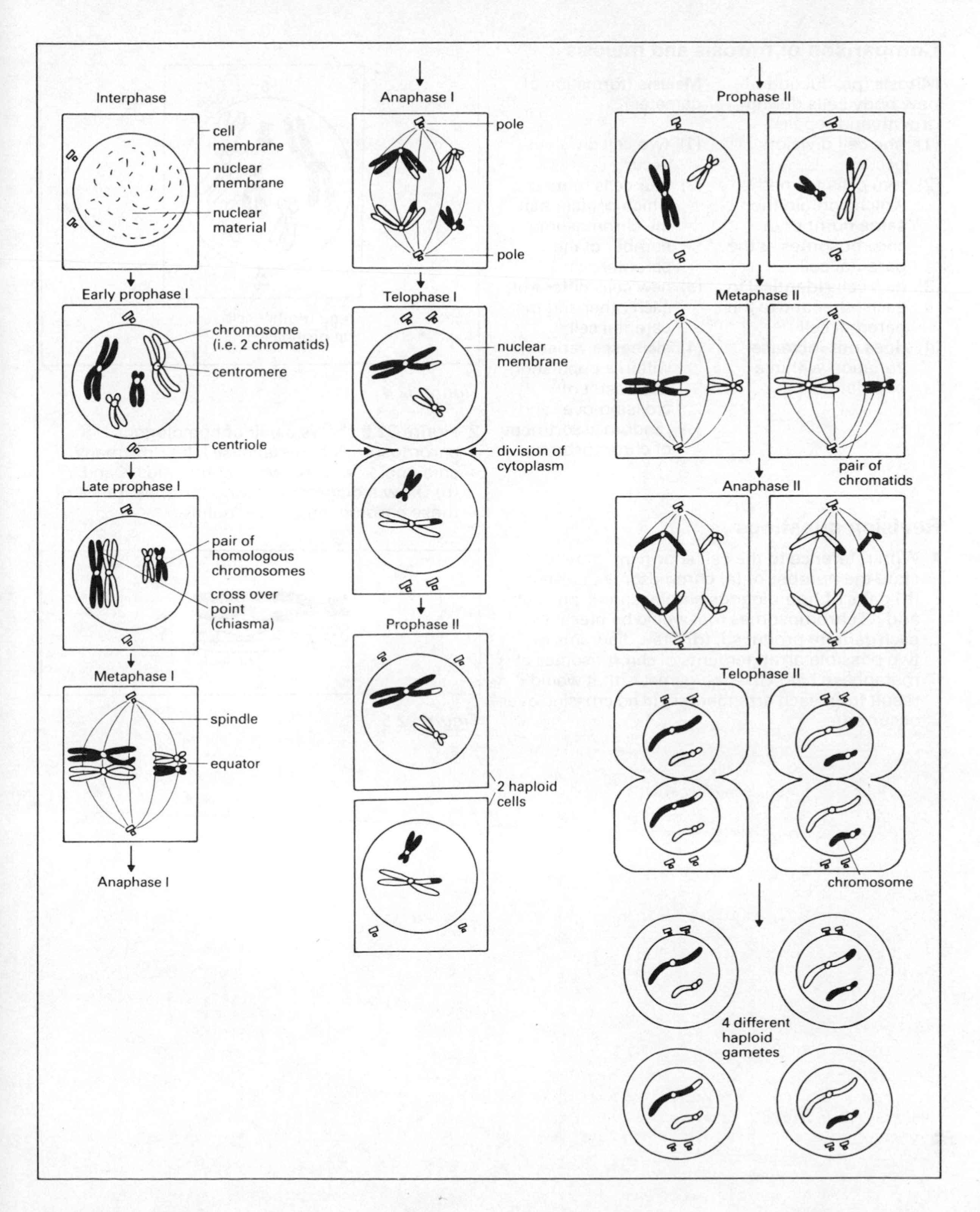

Figure 32.2 Meiosis

Comparison of mitosis and meiosis

Mitosis (production of new body cells during growth and repair)	**Meiosis** (formation of gametes)
(1) **one** cell division occurs	(1) **two** cell divisions occur
(2) **two** cells formed which contain the **same** number of chromosomes as the parental cell	(2) **four** cells formed which contain **half** the chromosome number of the parental cell
(3) new cells **identical** to each other and to the parental cell	(3) new cells **differ** from each other and the parental cell
(4) **does not increase** variation within a population	(4) **increases** variation within a population as a result of crossing over and random assortment of chromosomes.

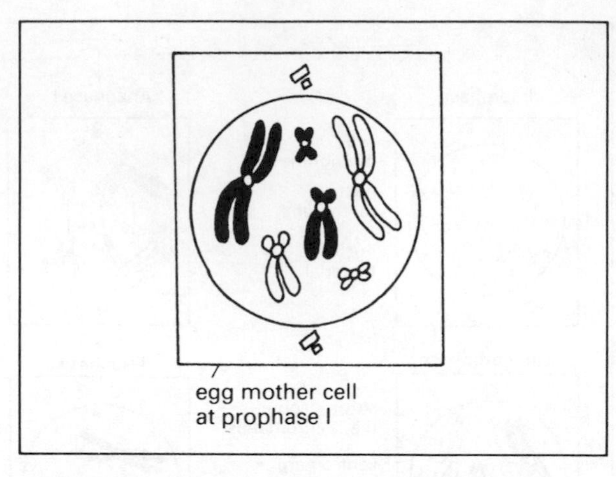

Figure 32.4

Revision questions

1 With reference to the cell shown in figure 32.4, state the number of (a) chromosomes present (b) pairs of homologous chromosomes present and (c) chromosomes that would be present in each gamete produced. (d) Draw diagrams of two possible arrangements of chromosomes at metaphase I. (e) Give the gametes that would result from each arrangement (if no crossing over occurred).

2 Figure 32.5 shows a pair of homologous chromosomes at metaphase I. (a) How many chiasmata occur between chromatids X and Y? (b) Draw a diagram showing the appearance of these chromosomes at telophase I.

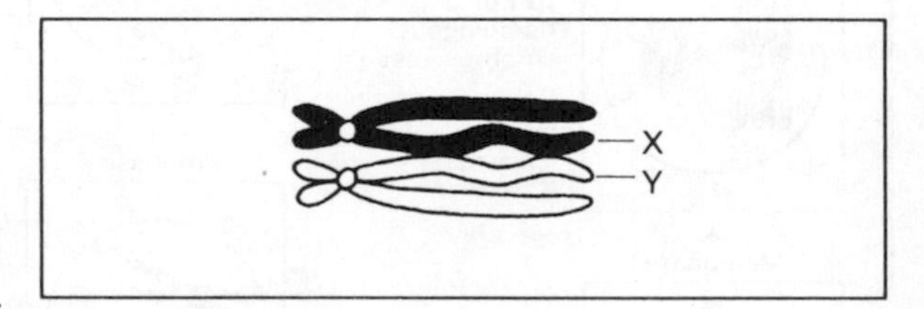

Figure 32.5

33 Monohybrid cross

Mendel (1822–1884), an Austrian monk, performed early experiments in genetics using different varieties of pea plant.

Mendel's first experiment

He crossed tall pea plants which were true-breeding (since when self-pollinated they always produced tall plants) with true-breeding short pea plants. He put pollen from the short type on to the stigmas of the tall type. All of the peas produced grew into tall plants. These are called the **first filial generation (F_1)**. He then self-pollinated these F_1 tall plants and the peas produced grew into the **second filial generation (F_2)**. Some of these plants were tall and some short in the ratio **3 tall:1 short.** This experiment is summarised as follows:

original cross	tall × short
F_1	all tall
second cross	tall × tall
F_2	3 tall:1 short

Short plants, absent in the F_1 generation, have reappeared in the F_2 so 'something' has been transmitted undetected in the gametes from generation to generation. Mendel called this a factor. To-day we call it a **gene.** The gene in the above cross controls the characteristic of height in pea plants. There are various forms **(alleles)** of this gene for height. One expression (allele) produces tall plants, another short plants. Since the presence of the tallness allele **masks** the presence of the shortness allele, the tallness allele is said to be **dominant** and the shortness one **recessive.**

Genes and chromosomes

Genes are carried on chromosomes. Each chromosome bears the same pattern of genes as its homologous partner. The position of a gene on a chromosome is called its **locus** (figure 33.1). Consider the gene for height. Let the allele for tallness be T and allele for shortness be t. Since each true-breeding tall plant has two tallness alleles, its genotype is TT and since each true-breeding short plant has two shortness alleles, its genotype is tt. At meiosis every gamete made by a TT plant receives one T allele and every gamete made by a tt plant receives one t allele. Therefore the cross can be represented as follows:

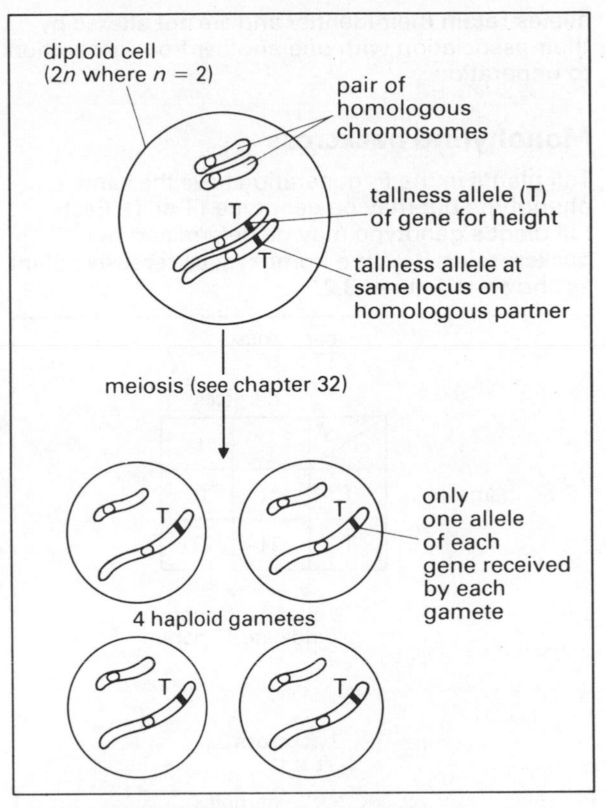

Figure 33.1 Genes and chromosomes

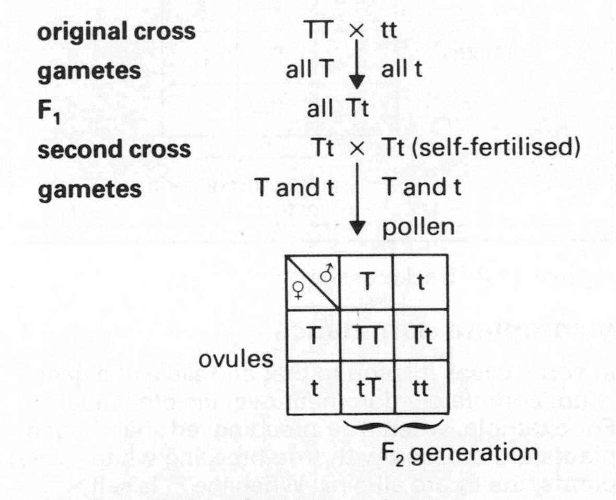

F_2 phenotypic ratio 3 tall : 1 short

The conclusions from Mendel's first experiment are expressed as Mendel's first law, the **principle of segregation,** and are as follows, using modern terms.

The alleles of a gene exist in pairs but when gametes are formed, the members of each pair pass into different gametes so that each gamete contains only one of the pair of alleles of the gene. The alleles retain their identity and are not altered by their association with one another from generation to generation.

Monohybrid backcross

Tall plants in the F_2 generation have the same phenotype but may be genotype TT or Tt. Each tall plant's genotype may be determined by **backcrossing** it with a **homozygous recessive** plant as shown in figure 33.2.

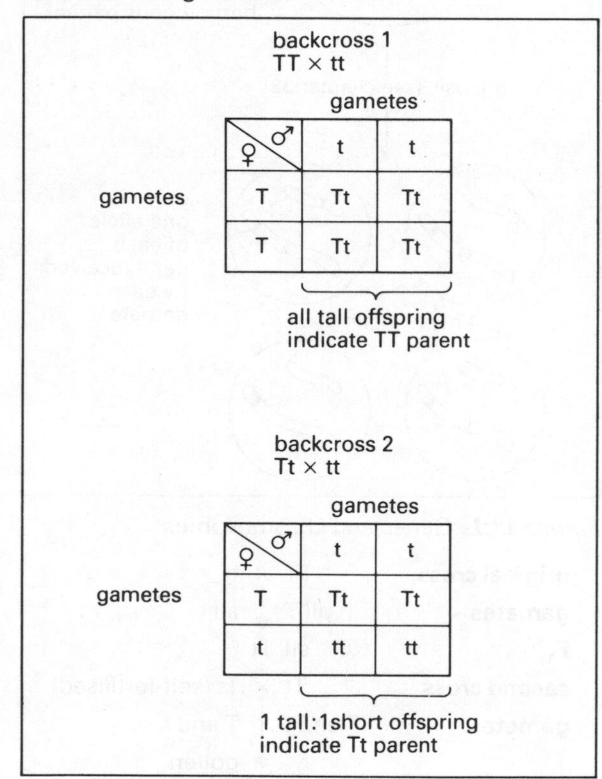

Figure 33.2 Backcrosses

Incomplete dominance

In some cases it is found that one allele of a gene is not completely dominant over the other allele. For example, when true-breeding red snapdragon plants are crossed with true-breeding white (ivory) plants, the F_1 are all pink. When the F_1 is self-pollinated, the F_2 occurs in the ratio 1 homozygous red: 2 heterozygous pink:1 homozygous ivory. This cross is represented as follows using R for the red allele and I for the ivory allele.

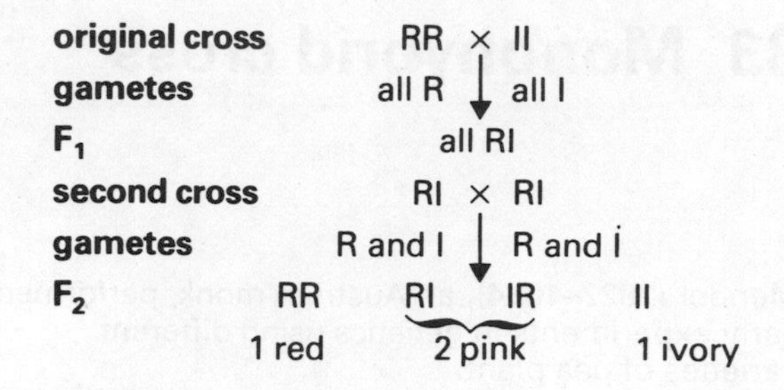

Multiple alleles

Sometimes more than two alleles of a gene exist. For example, one of the genes that affects coat colour in guinea pigs has four alleles which show complete dominance in the order C (black) dominant to C^k (brown) dominant to C^d (cream) dominant to C^a (albino).

Revision questions

1 Give a genotype for the following cattle where black is dominant to white, (a) a true-breeding black cow (b) a homozygous recessive cow and (c) a heterozygous bull.

2 In *Drosophila,* the allele + (normal wing) is dominant to vg (vestigial wing). Follow, in diagrammatic form, crosses (a) and (b) through to the F_2 generation giving the phenotypes of the offspring formed in each generation.
(a) + vg × vg vg (b) + + × + vg.

3 A self-coloured rabbit was crossed with a homozygous patch-coloured rabbit. Interbreeding amongst the F_1 generation produced an F_2 generation of 29 patch-coloured and 11 self-coloured rabbits. How could the genotype of one of the F_2 patch-coloured rabbits be determined?

34 Dihybrid cross

Mendel's second experiment

This cross involved plants differing in two characteristics, shape and colour. Mendel crossed true-breeding pea plants which produced **round yellow** peas with true-breeding plants which produced **wrinkled green** peas. All the F_1 plants produced round yellow peas showing round to be dominant to wrinkled and yellow dominant to green. This cross followed through to the F_2 generation is shown in figure 34.1 where R and r are the round and wrinkled alleles of the shape gene and Y and y are the yellow and green alleles of the colour gene. The **9:3:3:1** ratio (obtained from the results of crossing hundreds of plants) was explained by Mendel in the following way:

During gamete formation, the two alleles of a gene segregate into different gametes independent of the segregtion of the two alleles of another gene. (This means that there is just as much chance of R and y ending up together in a gamete as R and Y going together. Similarly r and Y end up together just as often as r and y). This is Mendel's second law, the **principle of independent assortment.**

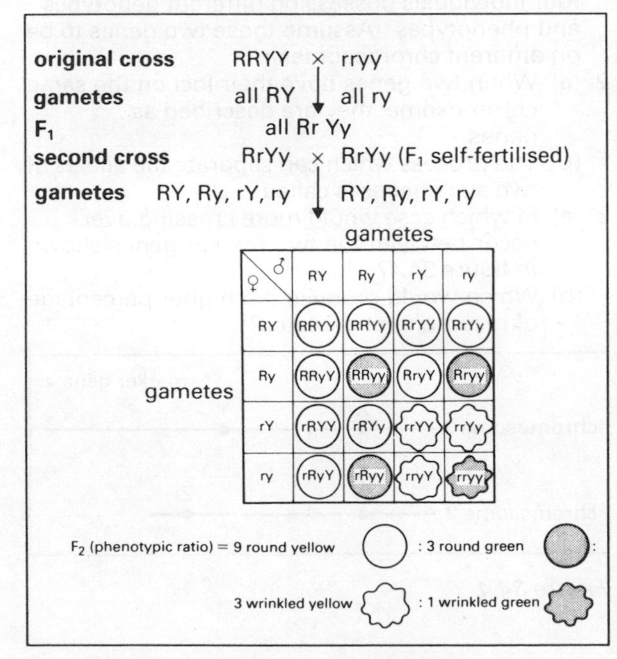

Figure 34.1 Mendel's second experiment

Gamete formation

Consider gamete formation in an F_1 individual (RrYy) assuming that the two genes involved are located on different chromosomes. As a result of **random assortment** of chromosomes during meiosis, there are two ways in which the homologous pairs can become arranged at metaphase I (figure 34.2). One arrangement produces RY and ry gametes, the other Ry and rY gametes. Therefore these four types of gametes are produced in equal numbers and each has an equal chance of being fertilised by any one of the other F_1's gametes. The sixteen possible combinations are shown in the punnet square in figure 34.1.

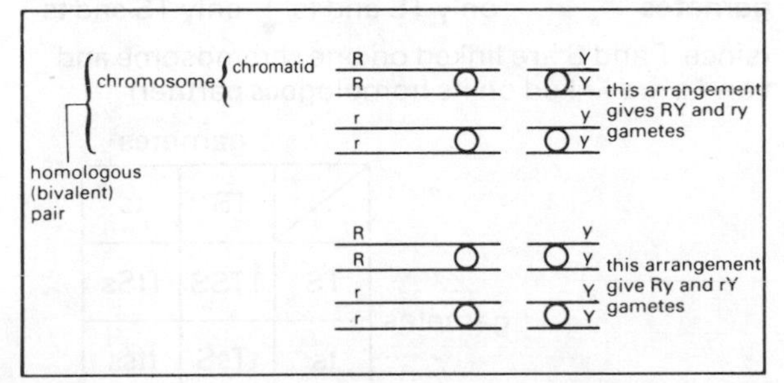

Figure 34.2 Formation of four gamete types

Dihybrid backcross

If the F_1 pea plants produced in the above example are backcrossed to a **double recessive** plant (instead of being self-fertilised), then the following occurs:

second cross RrYy × rryy

gametes RY,Ry,rY,ry all ry

F_2 genotypes RrYy Rryy rrYy rryy

phenotypes round, yellow; round, green; wrinkled, yellow; wrinkled, green

ratio 1 : 1 : 1 : 1

Conclusions

In a dihybrid cross where the genes show independent assortment, **self-fertilising** the F_1 results in an F_2 with the phenotypic ratio **9:3:3:1** whereas backcrossing the F_1 results in an F_2 with the phenotypic ratio **1:1:1:1**.

Linkage

Since there are many more genes than chromosomes, each chromosome must carry many genes. Thus when a cross is made which involves two alleles of two different genes located on the same chromosome, the two genes are transmitted together at meiosis and do not show normal segregation. Such genes on the same chromosome are said to be **linked.**

Consider the following example for tomato plants involving two linked genes, the gene for **height** (alleles T for tall, t for dwarf) and the gene for **stem type** (alleles S for smooth and s for hairy). If linkage between the two genes was complete then the following results would be obtained:

original cross TTSS × ttss

gametes all TS ↓ all ts

F_1 all TtSs

second cross TtSs × TtSs (F_1 selfed)

gametes only TS and ts ↓ only TS and ts

(since T and S are linked on one chromosome and t and s are linked on its homologous partner)

gametes

♀ \ ♂	TS	ts
TS	TTSS	TtSs
ts	tTsS	ttss

gametes

F_2 phenotypic ratio 3 tall smooth : 1 dwarf hairy

However **complete** linkage between two genes is **rare.** In the above example a small proportion of the F_2 generation always show random assortment (i.e. a few tall hairy and a few dwarf smooth plants appear).

Mechanism by which linked genes are separated
The **recombinants** (new forms) present in the F_2 appear as a result of crossing over occurring during meiosis at chiasmata between the two linked genes on the chromosome forming a few recombinant gametes e.g. Ts and tS as shown in figure 34.3.

Since chiasmata occur at any point on a chromosome, more crossing over (and therefore more recombination) occurs between two distantly located genes than two that are close together.

Thus an indication of the distance between genes on a chromosome is given by the percentage of recombination occurring between them. The greater this percentage, the greater the distance between the two genes. This can be used to map the order and location of genes on a chromosome.

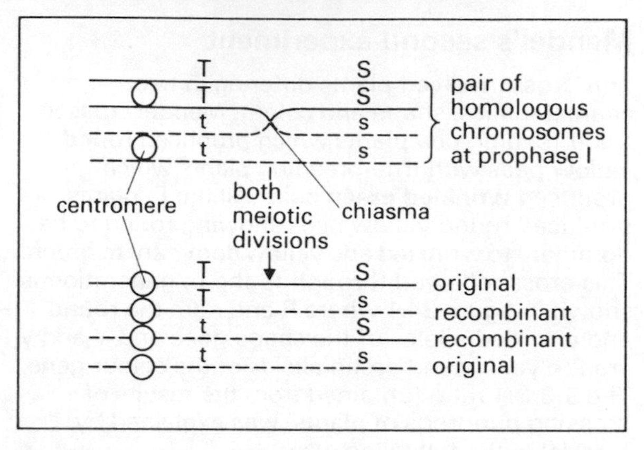

Figure 34.3 Effect of crossing over

Revision questions

1 In diagrammatic form, follow to the F_2 generation a cross between an ebony-bodied (e), vestigial-winged (vg) fruit fly and a true-breeding wild type (where + represents all dominant wild type characteristics). In your punnet square, underline four individuals possessing different genotypes and phenotypes. (Assume these two genes to be on different chromosomes).

2 (a) When two genes have their loci on the same chromosome, they are described as . . . genes.

(b) The process which can separate the alleles of two such genes is called . . . ..

3 (a) In which case would more crossing over occur between the two marker genes shown in figure 34.4?

(b) Which would produce the higher percentage of recombinant gametes?

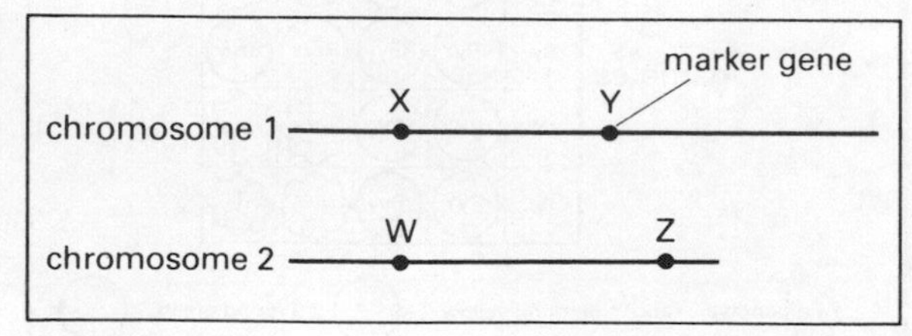

Figure 34.4

35 Sex determination and sex linkage

Sex determination

In the nucleus of every normal human somatic (body) cell, there are **46** chromosomes as **22** homologous pairs of **autosomes** and one pair of **sex** chromosomes. This latter pair determine an individual's sex.

In the female, the sex chromosomes make up a homologous pair, the **X** chromosomes. Thus a female has the chromosome complement, **44** + **XX**.

In the male, the sex chromosomes are an unmatched pair, an **X** chromosome and a smaller **Y** chromosome. Thus a male has the chromosome complement, **44** + **XY**.

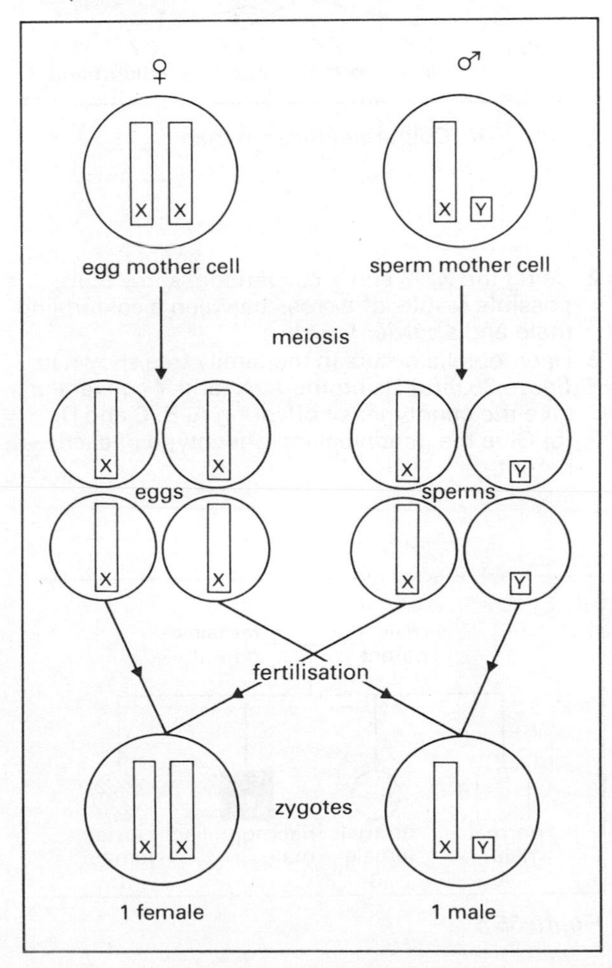

Figure 35.1 Sex determination in man (autosomes not shown)

An egg mother cell (figure 35.1) is **homogametic** since every egg formed by meiosis contains an X chromosome. A sperm mother cell is **heterogametic** since half of the sperms formed contain an X chromosome and half a Y chromosome. When the nucleus of a sperm fuses with the nucleus of an egg (ovum) at fertilisation (figure 35.1), the sex of the zygote formed is determined by the type of sex chromosome carried by the sperm.

In the locust, the female has an even number of chromosomes, 18 (16 + XX) and the male has an odd number, 17 (16 + X). Thus half of the sperms do receive an X chromosome, the other half do not and sex is determined as before.

The female is not always the homogametic sex. For example, in birds the female is heterogametic (XY) and the male is homogametic (XX).

Sex linkage

The smaller Y chromosome in man is homologous with only a short length of the longer X chromosome. The X chromosome therefore carries many genes not present on the Y. **Sex-linked genes** (figure 35.2) occur on sex chromosomes. When a sex-linked gene occurs on the X chromosome but not on the Y and that X chromosome meets a Y at fertilisation, then the sex-linked characteristic, whether dominant or recessive, is expressed in the phenotype. This is because the Y chromosome has no allele at the equivalent locus to offer dominance.

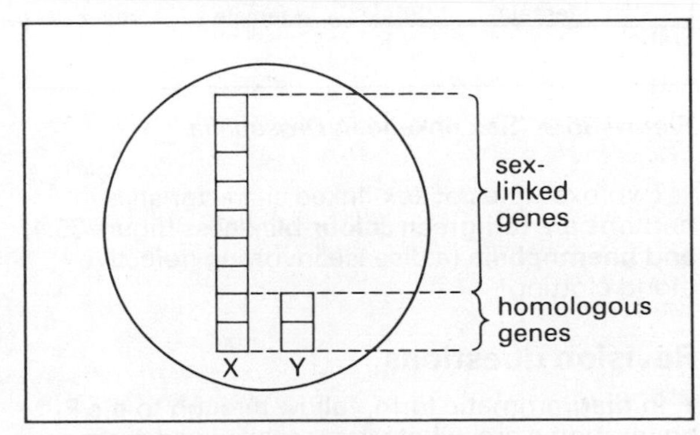

Figure 35.2 Sex-linked genes

Eye colour in Drosophila

Drosophila possesses the same mechanism of sex determination as man. The gene for eye colour occurs on the X chromosome as a sex-linked gene. The allele for red eye colour **(R)** is dominant to the allele for white **(r)**. Figure 35.3 shows a cross between a white-eyed female and a red-eyed male followed through to the F_2 generation. Because a sex-linked gene is involved, the normal 3:1 ratio found in the F_2 of a monohybrid cross does not occur.

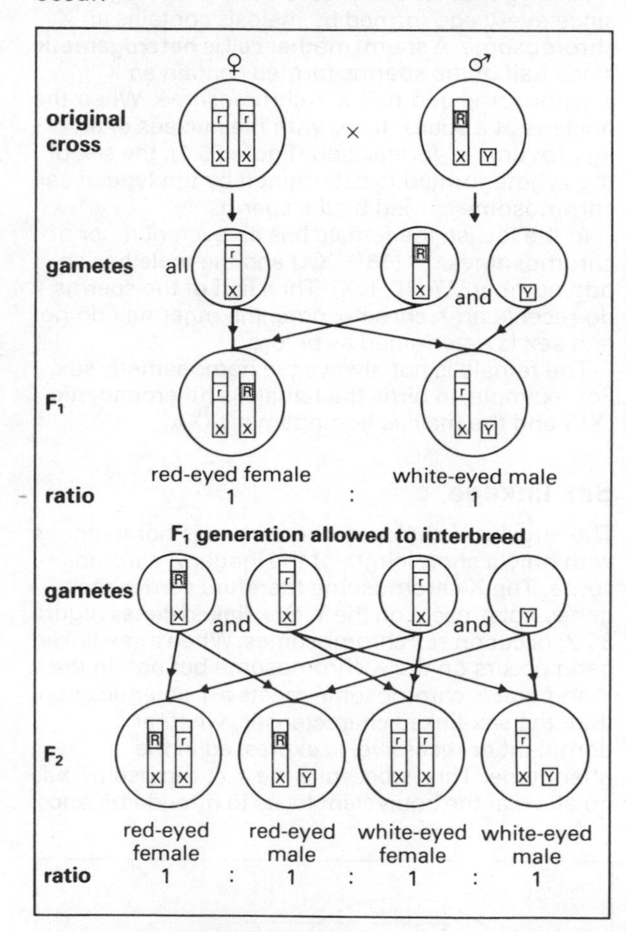

Figure 35.3 Sex linkage in *Drosophila*

Two examples of sex-linked characteristics in humans are **red-green colour-blindness** (figure 35.4) and **haemophilia** (a disease involving defective blood clotting).

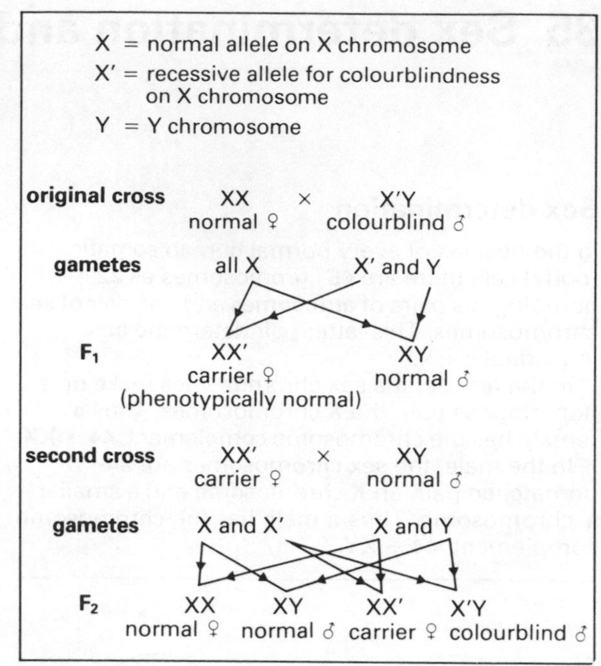

Figure 35.4 Colourblindness in man

Revision questions

1 In diagrammatic form, follow through to the F_2 generation a cross between a white-eyed male *Drosophila* and a homozygous red-eyed female, where the F, are allowed to interbreed.

2 Using the X, X′ and Y convention, show the possible results of a cross between a colourblind male and a carrier female.

3 Haemophilia occurs in the family tree shown in figure 35.5 (a) Using the X, X′ and Y convention, give the genotypes of offspring A, B, C and D. (b) Give the genotype and phenotype of each parent.

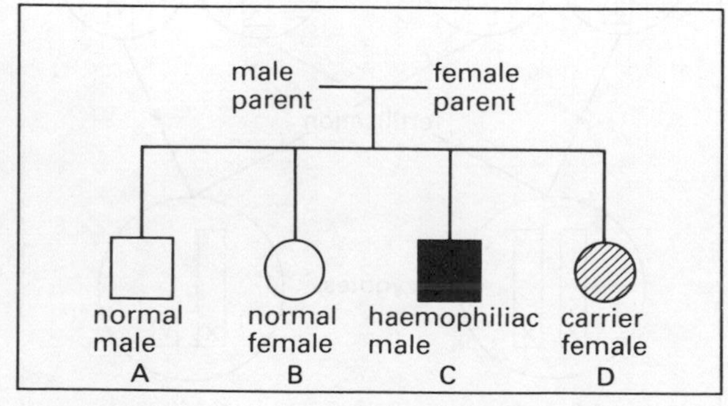

Figure 35.5

36 Mutation

A **mutation** is a change in the genotype of an individual. When this produces a change in the phenotype, the individual affected is called a **mutant.**

Gene mutations

A **gene mutation** involves a change in one or more **nucleotides.** In each of the examples shown in figure 36.1, the codon for a particular amino acid is altered,

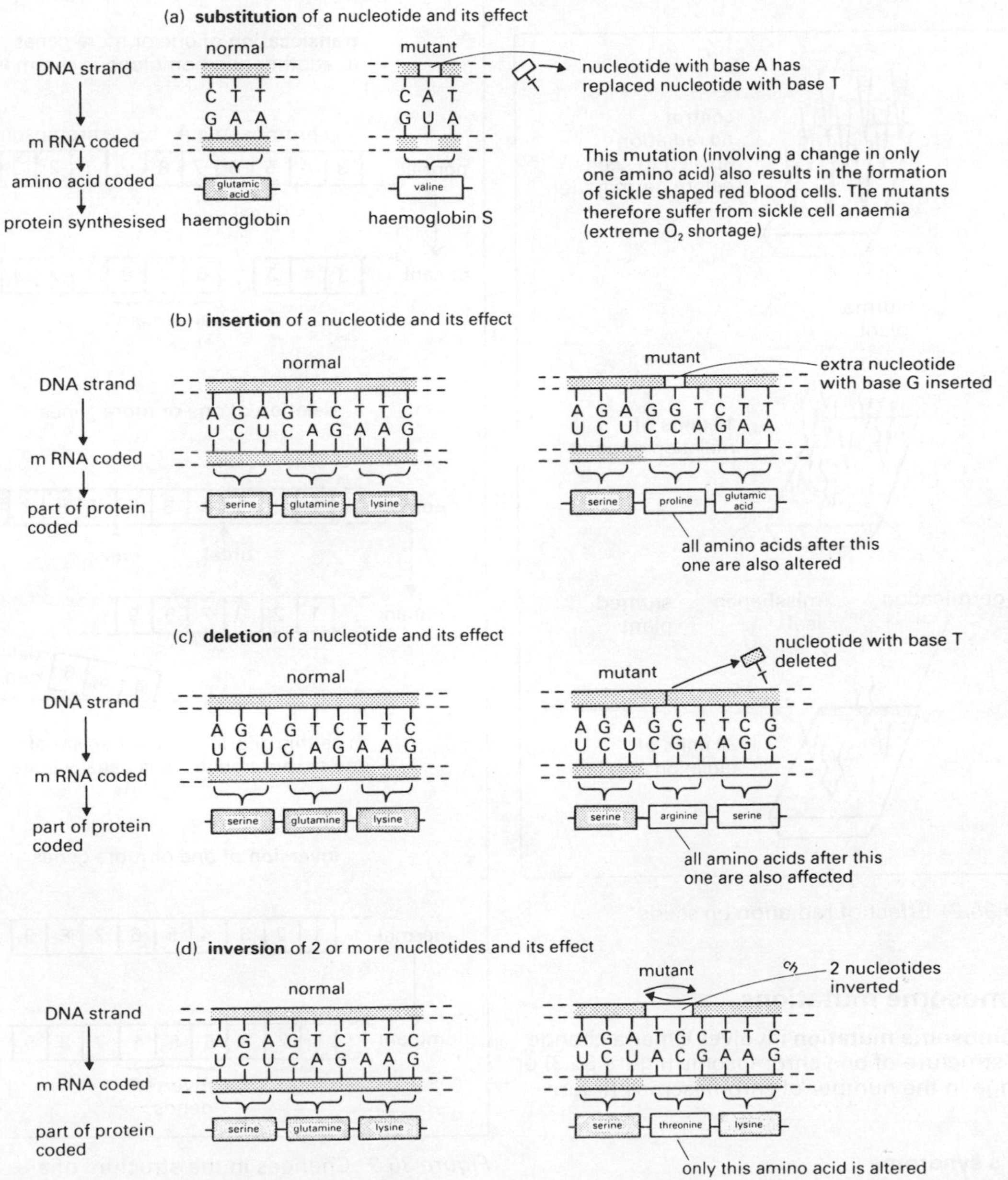

Figure 36.1 Types of gene mutation

changing the protein that is synthesised. If an essential protein is affected and death results, then the altered gene is known as a **lethal** gene.

Gene mutations occur spontaneously and can be induced by **mutagenic agents** such as high temperature, nitrogen-mustard gas and various types of radiation (atomic, UV light and X-rays). The effect of radiation on seeds is shown in figure 36.2. From this experiment it can be concluded that an increase in level of radiation brings about an increase in rate of gene mutation.

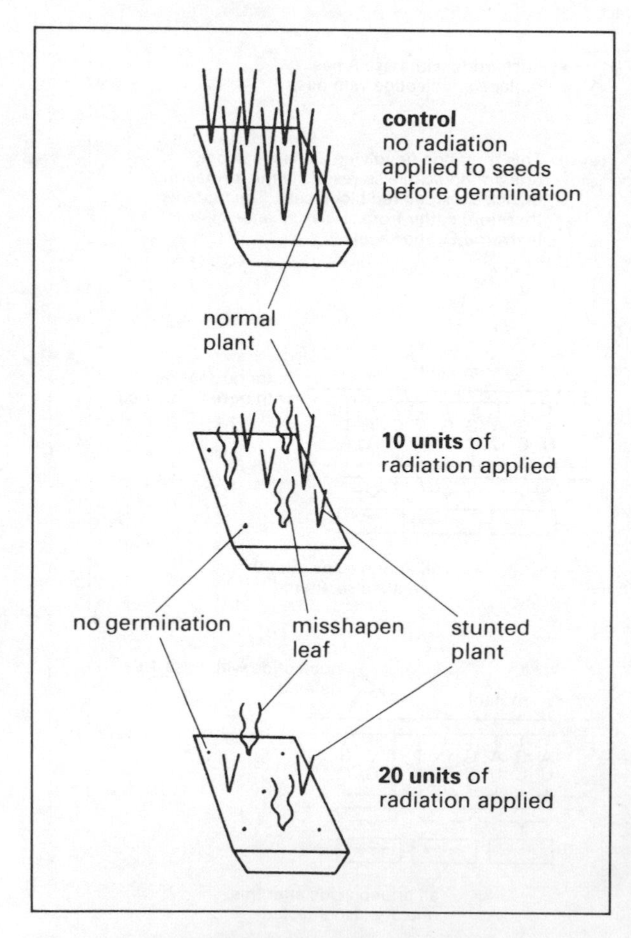

Figure 36.2 Effect of radiation on seeds

Chromosome mutations

A **chromosome mutation** involves either a change in the structure of one chromosome (figure 36.3) or a change in the number of chromosomes (figure 36.4).

Down's syndrome

If **non-disjunction** (figure 36.4b) occurs during

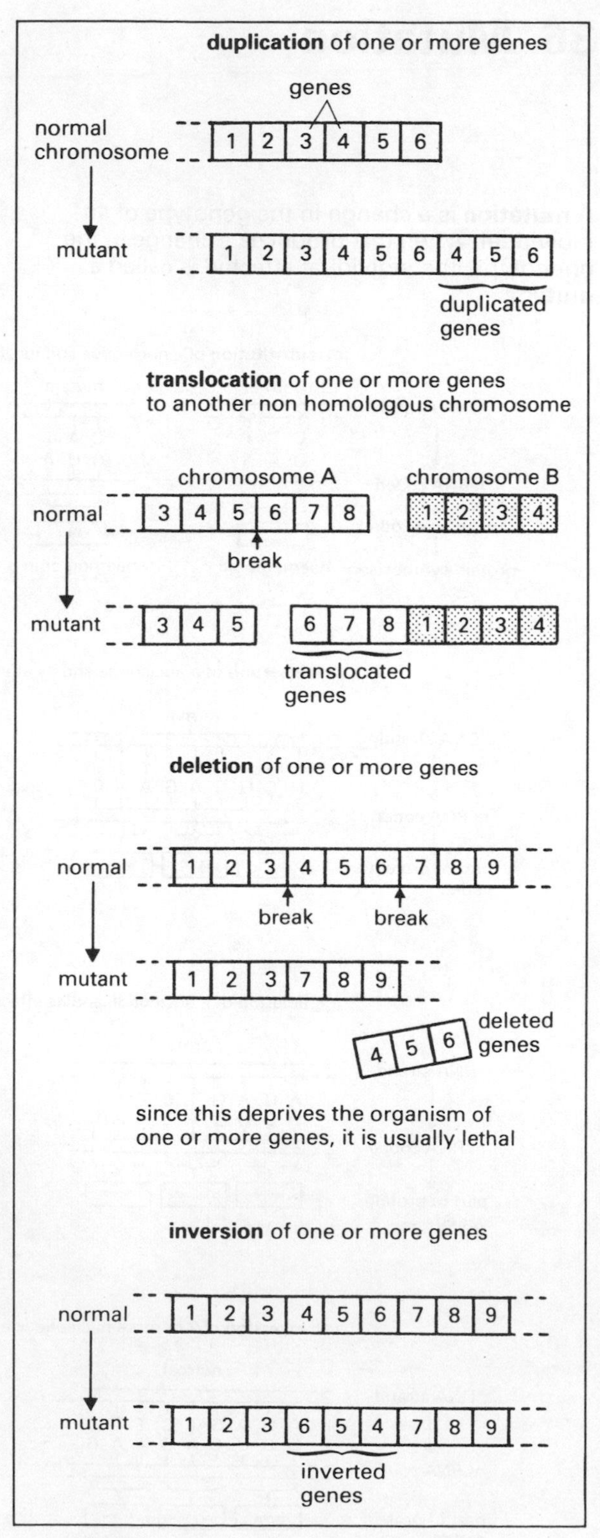

Figure 36.3 Changes in the structure of a chromosome

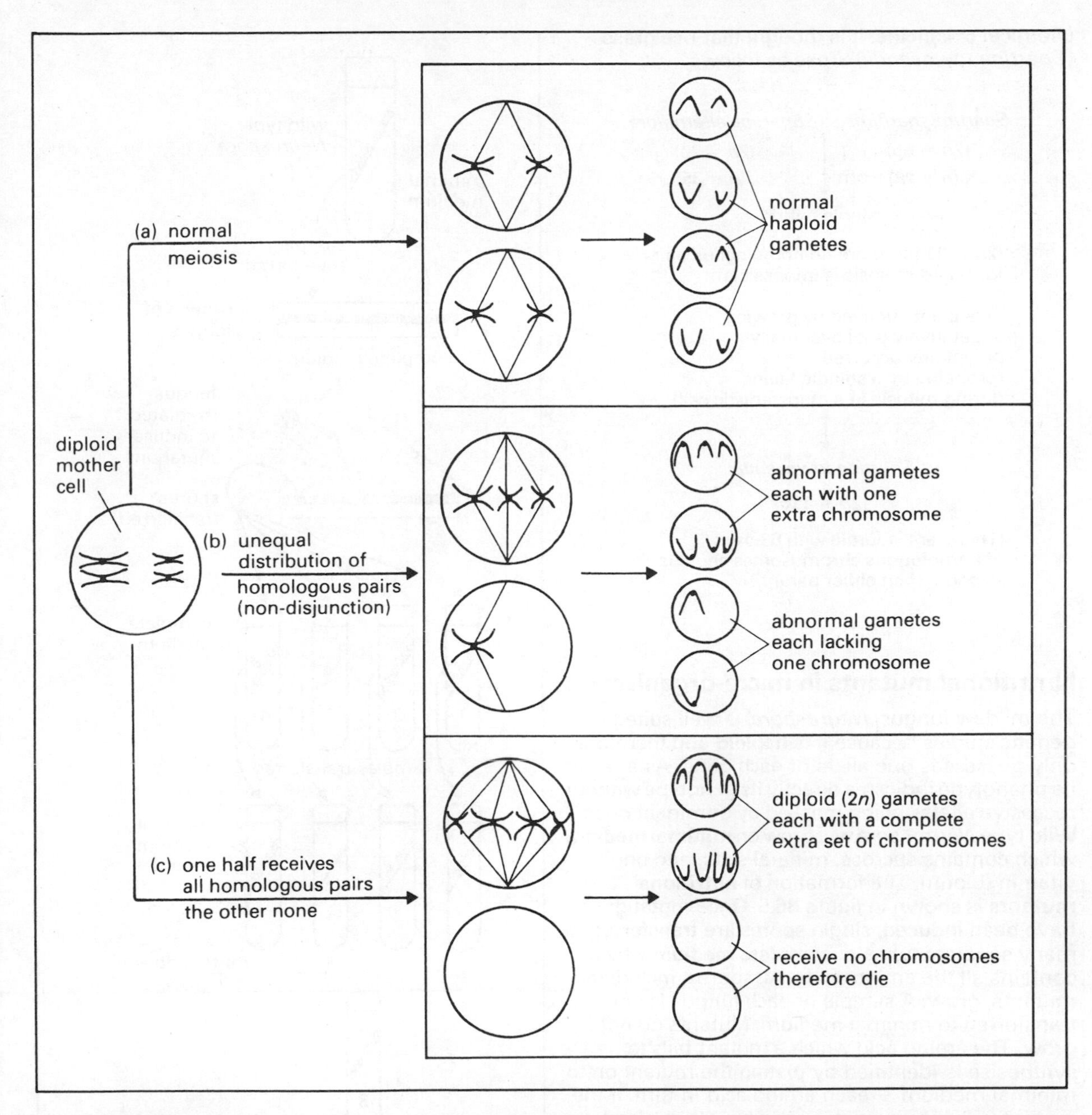

Figure 36.4 Changes in the number of chromosomes

human gamete formation and an abnormal egg ($n = 24$) is fertilised by a normal sperm ($n = 23$) then the resultant zygote is $2n = 47$. Such an individual suffers from **Down's syndrome** (mongolism).

Polyploidy

Polyploidy is an increase in the number of all of the chromosomes by three, four or more times the normal haploid count. This occurs when a spindle fails during mitosis or when all the chromosomes undergo **non-disjunction** simultaneously (figure 36.4c). Fusion of a diploid gamete ($2n$) and a normal haploid gamete (n) produces a triploid ($3n$) zygote. This is sterile since it cannot make homologous pairs of chromosomes at meiosis. Fusion of two diploid gametes produces a tetraploid ($4n$) zygote.

Polyploid plants usually show an increase in size and resistance to disease. Many crop plants are polyploid and therefore give a bigger yield than their diploid relatives. Polyploidy can be induced by the chemical **colchicine.** It is thought that rice grass (*Spartina townsendii*) arose as follows:

Spartina maritima × *Spartina alterniflora*
($2n = 56$) ($2n = 70$)
($n = 28$) ($n = 35$)
↓
sterile hybrid

($2n = 63$ therefore no homologous pairs form and meiosis is impossible)

This plant survived by growing vegetatively until eventually polyploidy occurred (probably by a spindle failing during mitosis in a meristematic cell)

↓

Spartina townsendii

($2n = 126$)

(This plant is fertile with 63 pairs of homologous chromosomes and it is stronger than either parent.)

Nutritional mutants in micro-organisms

The mildew fungus, *Neurospora,* is well suited for genetic studies because it is **haploid** and therefore only possesses one allele of each gene. As a result its phenotype indicates directly its genotype without recessive alleles being masked by dominant ones. Wild type *Neurospora* will grow on **minimal medium** which contains sucrose, mineral salts, and one vitamin (biotin). The formation of **nutritional mutants** is shown in figure 36.5. Once mutations have been induced, single spores are transferred to many separate tubes of **complete medium** which contains all the amino acids. All spores, including mutants, grow. A sample of each fungus is next transferred to minimal medium. Mutants do not grow. The amino acid which a mutant fails to synthesise is identified by plating the mutant on to minimal medium + each amino acid in turn. If the mutant will only grow, for example, on minimal medium + arginine, then it is an arginine-requiring strain of *Neurospora.*

When many nutritional mutants are produced in this way, it is found that several different mutant strains will grow if given arginine. Table 36.1 shows that of these, strain 1 will only grow if supplied with arginine, whereas strain 2 will grow if supplied with citrulline or arginine and strain 3 will grow if supplied with ornithine, citrulline or arginine.

Complex chemicals are synthesized in a cell by a series of reactions, for example W → X → Y → Z. Each stage in the series is controlled by an enzyme.

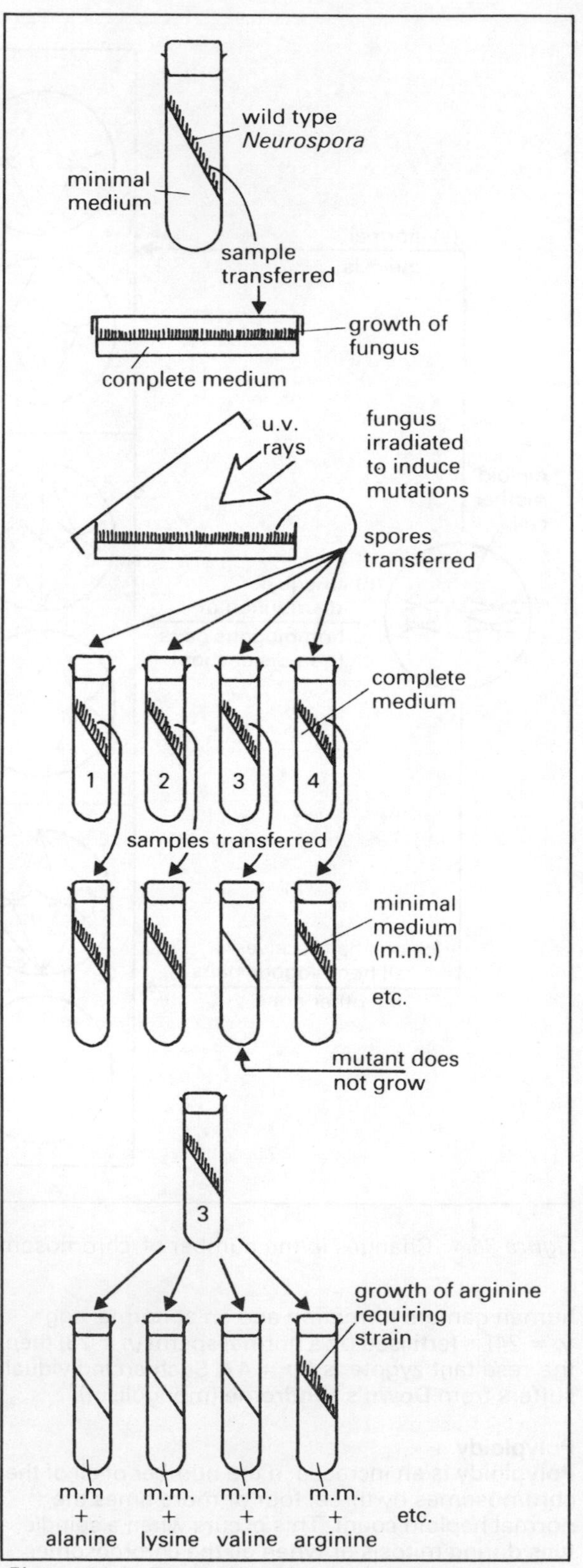

Figure 36.5 Formation of nutritional mutants of *Neurospora*

Mutant strain	Minimal medium alone	m.m. + ornithine	m.m. + citrulline	m.m. + arginine
1	–	–	–	+
2	–	–	+	+
3	–	+	+	+

+ = growth – = no growth

Table 36.1 Nutritional mutants of *Neurospora*

Therefore in the building up of arginine from other organic compounds, the last three stages in the chain are as follows:

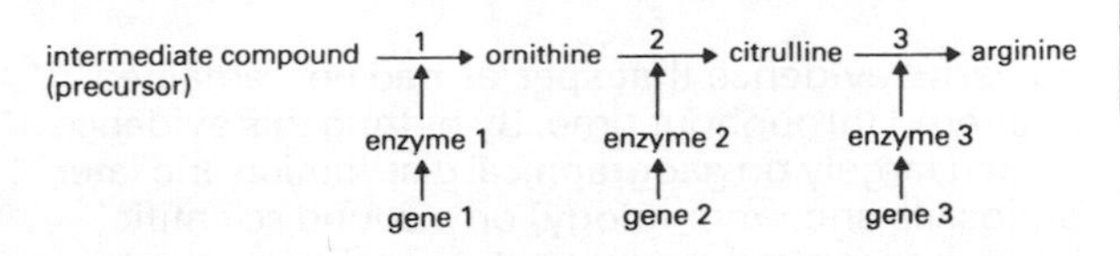

Thus an arginine-requiring strain has its metabolic block at position 3 since it cannot make enzyme 3. It accumulates citrulline as a result of its metabolism and will only grow if supplied with arginine. A citrulline-requiring strain has its metabolic block at position 2 since it cannot make enzyme 2. It accumulates ornithine and grows if given citrulline or arginine. An ornithine-requiring strain has its metabolic block at position 1 since it cannot make enzyme 1. It accumulates the precursor before ornithine and grows if supplied with ornithine, citrulline or arginine.

It is thought that a gene controls the production of each enzyme in the pathway. If X-rays cause a gene to mutate, it can no longer code for its particular enzyme and a metabolic block occurs. This 'one gene—one enzyme' hypothesis has been demonstrated in many organisms.

Mutation as a source of variation

Most mutations are harmful or even lethal. However on very rare occasions, there occurs by mutation a mutant gene which confers some advantage on the organism and is therefore better than the original gene. Such mutant genes provide the alternative choices upon which natural selection can act and they are therefore considered to be the raw material of evolution (see chapters 37 and 38).

Revision questions

1 In the following two telegrams, a small error alters the sense of the message.
(a) Intended: John walked to the bus.
Actual: John talked to the bus.
(b) Intended: The able should walk.
Actual: The bale should walk.
To which type of gene mutation is each of these analogous?

2 The homologous pair of chromosomes shown in figure 36.6 was seen under the microscope during meiosis in a cell. Which type of chromosome mutation had occurred?

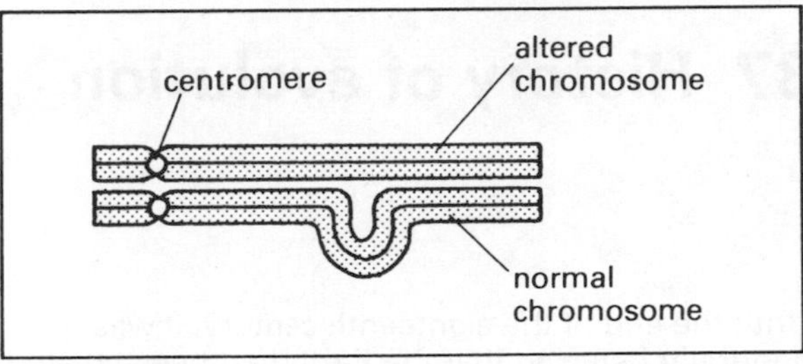

Figure 36.6

3 The following cross shows the genetic relationships between some species of *Primula:*

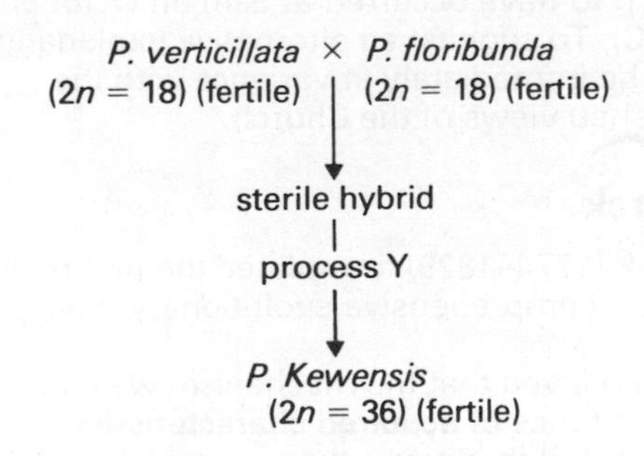

(a) What is the diploid ($2n$) chromosome complement for the sterile hybrid?
(b) Explain why this hybrid is sterile.
(c) Name process Y.

4 In man the following series of reactions occurs during normal cell metabolism:

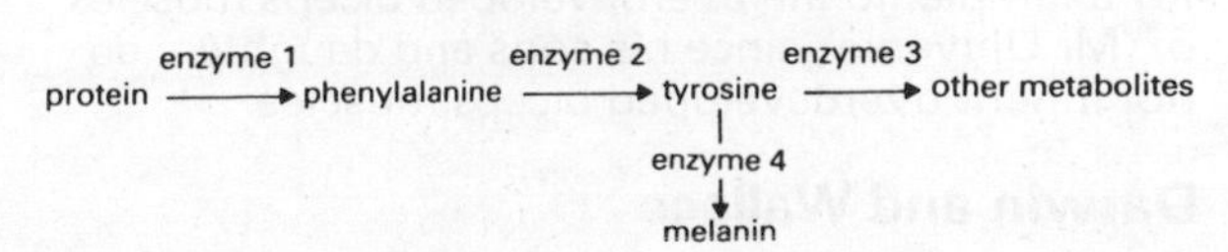

Pale skin and excess phenylalanine in the blood are symptoms of the hereditary disease, phenylketonuria. What mutation may have caused this disease?

5 Three mutants of *Neurospora* which fail to produce methionine were grown as shown in table 36.2. (a) Write down the correct sequence of these three metabolites as they occur in the synthesis of methionine. (b) Which gene has been altered in each mutant strain?

Mutant strain	Minimal medium alone	m.m. + methionine	m.m. + homocysteine	m.m. + cystathionine
1	–	+	+	+
2	–	+	+	–
3	–	+	–	–

Table 36.2

37 History of evolution

Until the end of the eighteenth century, it was generally believed that the Earth had been created in six days 'when certain elemental atoms were suddenly commanded, by an act of special creation, to flash into living tissues and form every species of animal and plant on Earth'. In addition, it was believed that every species had remained unaltered since that one grand moment in time (estimated by a bishop to have occurred at 9am on October 23rd, 4004 BC). To suggest an alternative explanation was to be immediately at variance with the established views of the Church.

Lamarck

Lamarck (1774–1829) formulated the first really clear and comprehensive evolutionary theory as follows:

(a) He believed that the mechanism was the inheritance of **acquired characteristics**.
(b) He stated that it was the action of the environment on the organisms that produced the change.
(c) In his view, this change in phenotype (brought about by the individual itself) was then passed on to the next generation.

The giraffe's neck is the classic example quoted (see figure 37.1). However, doubt is cast on the explanatory power of the theory when it is applied, for example, to the overdeveloped biceps muscles of 'Mr Universe', since his sons and daughters do not inherit overdeveloped biceps muscles.

Darwin and Wallace

Darwin (1809–1882) did two things. He presented powerful evidence that species had not remained unaltered throughout time. By putting this evidence (based largely on geographical distribution and later on fossils and embryology) on a sound scientific basis, he inspired a new attitude and approach to biological enquiry. His second contribution was to put forward a plausible hypothesis explaining the mechanism of evolution. This was published in 1859 as a joint paper with **Wallace.** In it, the authors stated that in their view, the main factor producing evolutionary change is **natural selection.**

A year later, Darwin amplified this view in his famous book *'The Origin of Species'* as follows:

(a) Organisms tend to produce more offspring than the environment will support.
(b) A struggle for existence follows and a large number of these offspring die before completing their life cycle due to overcrowding, lack of food and competition.
(c) Members of the same species are not identical but show variation in all characteristics.
(d) Those offspring whose phenotypes are better suited to their immediate environment will, in time, have a better chance of being among the survivors and will reach reproductive age and pass on the favourable characteristics to their offspring.
(e) Those offspring whose phenotypes are less well suited to their immediate environment will die before reaching reproductive age.
(f) Over a period of time, the best suited phenotypes will predominate in the population i.e. the fittest will survive.

Darwin called this 'weeding out' process natural selection and since environmental conditions are

Lamarck's theory

Short-necked ancestral giraffes had to stretch their necks in order to reach the leaves of the trees on which they fed.

Their offspring inherited these slightly longer necks which they in turn stretched during feeding.

This process was repeated many times until the long necks found in modern giraffes finally developed.

Darwin's theory

Since variation exists amongst the members of a species, some ancestral giraffes just happened to have slightly longer necks.

By natural selection, the longer necked giraffes survived since they could reach food and the short-necked ones died.

Only those animals with the longest necks survived and produced more long-necked giraffes.

Figure 37.1 Comparison of two theories of evolution

constantly changing, natural selection is for ever favouring the emergence of new forms.

Darwin formulated his theory without genetic background since Mendel's laws, although published in 1865, lay unnoticed until 1900. Had Darwin known the basis of genetics, it would probably have made a great difference to the development of his theory. He always felt the weakest point of his argument lay in the lack of an adequate explanation of how variation arose and was then transmitted from generation to generation.

Comparison of Darwin's and Lamarck's theories

As an example consider the giraffe. Fossil evidence shows that its early ancestors had short necks but that during geological history the neck became longer and longer. Figure 37.1 applies each of the two conflicting theories to this example of evolution. Much evidence (particularly genetic) is now available to support Darwin's theory but nothing has been found to support Lamarck.

Revision questions

1 Name Darwin's famous book on evolution.
2 Name Darwin's co-author of the paper which preceded that book.
3 (a) Whose theory of evolution is based on 'the inheritance of acquired characteristics'?
(b) Whose theory of evolution is based on 'survival of the fittest by natural selection'?
(c) How would each of these scientists have explained the evolution of flippers from legs in a certain species of water living mammal?

38 Mechanism of evolution

Role of variation

In an environment, the conditions are constantly changing. Since variation exists amongst the members of a species, some individuals are better suited to new conditions that arise and therefore survive by natural selection. The less well suited individuals fail to survive.

Sources of heritable variation

Random assortment of chromosomes during meiosis

Depending upon which way up the homologous chromosomes become attached at the equator, a **reshuffle** of genes results giving new combinations (see chapters 32 and 34). There are 2^{23} (8 388 608) possible chromosome combinations in man's gametes. This is one of the reasons that two brothers are never identical.

Crossing over

When this occurs at **chiasmata** between two chromatids at meiosis, new combinations of genes are established (see chapters 32 and 34).

Mutations

Chapter 36 explains how gene and chromosome mutations arise. Gene mutations are considered to be the more important type in evolution since these provide an enormous **reservoir** of alleles which can then be combined in various new ways by the above two processes. This ensures that variation exists amongst the members of a species.

Multiple alleles

Sometimes more than two alleles of a gene exist (e.g. coat colour in guinea pigs, chapter 33). Clearly however only two can occupy positions on a pair of homologous chromosomes at one time. A further example is the blood group system in man which is controlled by three alleles. A and B are equally dominant, O is recessive. The relationship between genotype and blood group is shown in table 38.1. Multiple alleles provide a source of variation since offspring can be produced that are different from either of the parents (figure 38.1).

Natural selection in action

Despite the fact that most mutations produce inferior versions of the original gene, in each of the

Genotype	Blood group
AA or AO	A
BB or BO	B
AB	AB
OO	O

Table 38.1 Blood groups

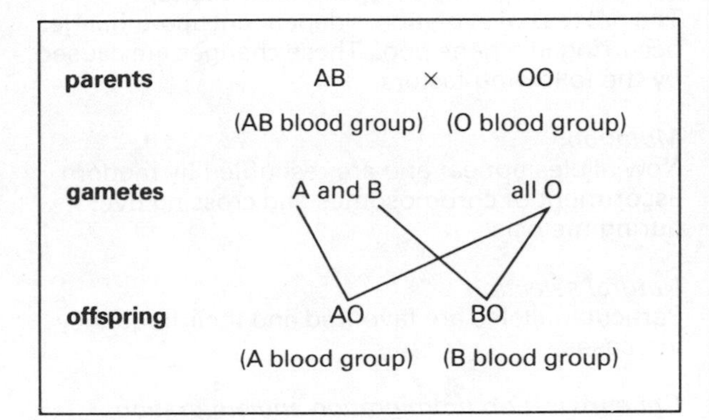

Figure 38.1 Effect of multiple alleles

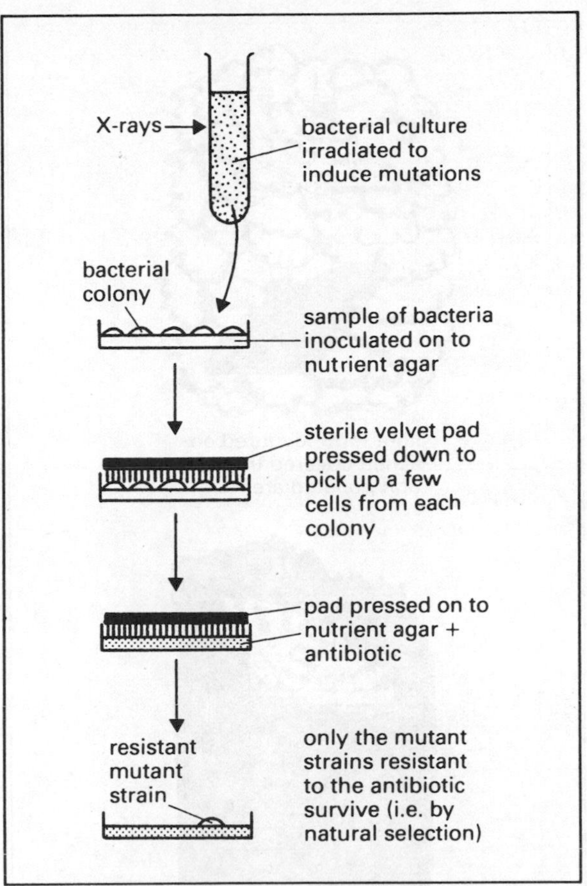

Figure 38.2 Isolation of resistant bacteria

following examples of populations, there occurs a mutant gene which allows adaptation to the changing environment. This gives the mutant forms of the organism a **selective advantage.**

Resistance to antibiotics

Some mutant forms of bacteria are resistant to antibiotics. The experiment shown in figure 38.2 demonstrates the occurrence of natural selection. Similarly when an antibiotic is injected into a diseased animal, any mutant bacteria that survive the antibiotic multiply since there is no competition. The disease may therefore continue.

Resistance to DDT

Some mutant forms of flies and mosquitoes are resistant to the pesticide DDT and therefore enjoy a selective advantage. If the use of DDT is continued, the 'mutants' increase in number replacing the 'wild type' which is destroyed by DDT.

Natural selection in the peppered moth (industrial melanism)

The two forms of this moth differ by only one **colour-determining gene**. One is pale with a light, speckled body, the other is the dark melanic form. In 1950, surveys showed that the pale form was most abundant in non industrial areas whereas the dark form was most common in areas suffering from heavy industrial air pollution. This moth flies by night and rests on the bark of trees during the day. In non polluted areas, the tree trunks are covered with pale-coloured lichens and the light coloured moth is well camouflaged against this pale background. However the dark form is easily seen and eaten by predators such as thrushes. In polluted areas, toxic gases and soot particles kill the lichens and darken the tree trunks. As a result, the light coloured moth is easily spotted whereas the dark one is well hidden (figure 38.3).

In polluted areas at present undergoing 'cleaning up campaigns', the pale form is being naturally selected at the expense of the melanic form which is losing its selective advantage in this changing environment.

Artificial selection in action

In animal and plant breeding, man selects the males and females with the characteristics that he wants and allows them to interbreed. Thus man is altering and in some cases speeding up the

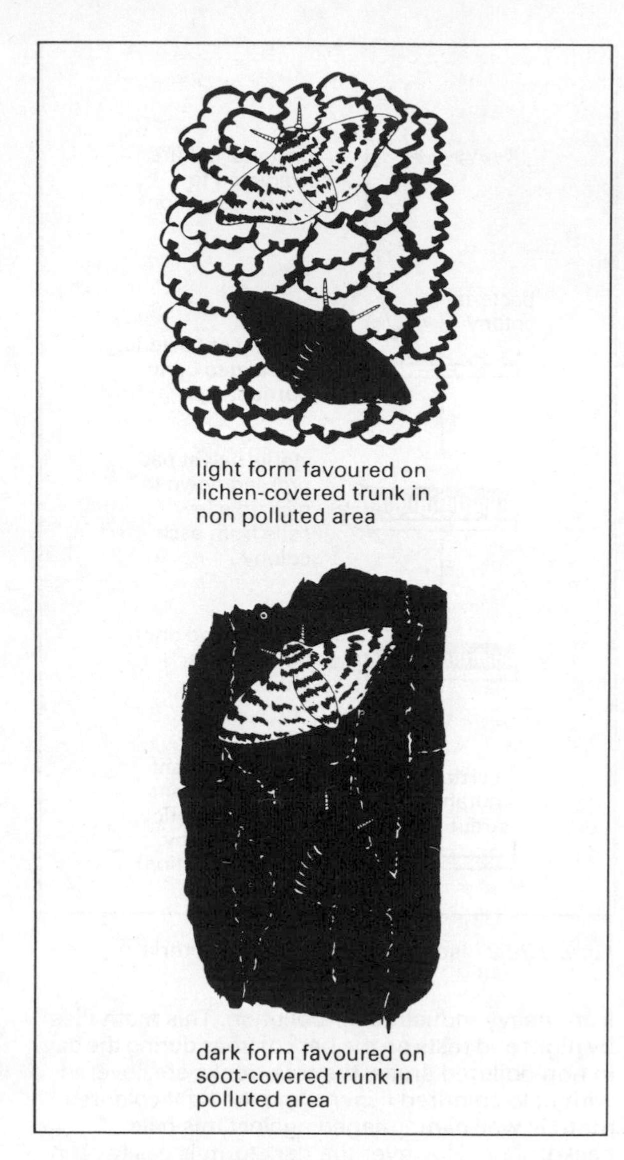

Figure 38.3 Natural selection in peppered moth

evolution of certain varieties of species as illustrated by the following examples. Jersey and Ayrshire cattle are bred for milk production and Hereford and Aberdeen Angus cattle are bred for meat production (see also chapter 39). Dogs are bred for hunting, racing and appearance. (The dramatic difference between the Chihuahua and the St Bernards provides powerful evidence of man's ability to alter a species.) Wheat, barley and potato are bred for higher yield and greater resistance to disease. (Wheat has been altered so much by artificial selection that it can no longer survive without man.)

Gene pool

The total of all the different genes in a population is known as the **gene pool.** The frequency of any gene in a population (relative to all of its alleles at the same locus) is known as the **gene frequency.** As long as the population is large, mating is random and none of the conditions listed below occurs, then the gene frequencies in the population remain constant from generation to generation and recessive alleles are not lost.

Alteration in gene pool (gene frequencies)
The process of evolution is dependent upon changes occurring in a gene pool. These changes are caused by the following factors.

Mutation
New alleles appear and are reshuffled by random assortment of chromosomes and crossing over during meiosis.

Natural selection
Particular alleles are favoured and their frequency increases.

Gene migration (immigration and emigration)
When immigrants arrive from another population possessing a different gene pool, new alleles are introduced.

Genetic drift
When a small group of organisms becomes isolated from the rest of the population, the small group often does not possess the complete range of genes typical of that species. After several generations, this isolated group becomes distinctive since the frequencies of certain genes have changed. For example America was first populated, it is thought, by a group of Asians who migrated across the ice packs of Bering Strait and became isolated from the rest of the Mongoloid race. Genetic drift accounts for the differences in the percentages of population possessing certain blood groups as shown in table 38.2.

People	% population with blood group			
	A	B	AB	O
Chinese	31	28	7	34
Sioux American Indians	7	2	0	91

Table 38.2 Effect of genetic drift

Speciation

Speciation is the **formation** of new species (usually many from a few). **Evolution** is the **mechanism** by which speciation is brought about and involves gradual changes in genotype (and phenotype) of a population. These changes are adaptive, making the organism better at exploiting the environment. Speciation occurs as follows:

Reduced selection pressure occurs and times become easy. For example, the population moves into a new environment free from competing species, predators, parasites etc. (When the common ancestor of Darwin's finches reached the Galapagos Islands, presumably it only found a few songbird species occurring there. If true woodpeckers and warblers had already been there, then the ecological niches would not have been available for woodpecker-like and warbler-like species of finch to evolve).

A **population explosion** follows which results in increased variation since many alleles, previously selected against, are now expressed.

In time the population expands its range and divides into two or more **sub-populations** that have little gene exchange.

Complete isolation of these sub-populations occurs, no interbreeding is possible and separate **subspecies** with separate gene pools result.

Such isolation is caused by the following **barriers** to gene exchange:

(a) **Geographical** e.g. water (oceans and rivers), mountain ranges and deserts.
(b) **Ecological** e.g. humidity and temperature.
(c) **Reproductive** e.g. lack of attraction between male and female, failure of courtship to stimulate partner, physical non correspondence of genital organs, failure of gametes to fuse, abortion of inferior zygote or production of sterile offspring (e.g. horse × donkey → sterile mule).

Over a very, very long period of time, the gene pools become so altered that the groups are genetically distinct i.e. **genetically isolated.**

On reunion, these groups can no longer interbreed (e.g. chromosomes cannot make homologous pairs etc.). Speciation has occurred and they are separate distinct **species.**

Revision questions

1 Why is the existence of variation within the members of a species insurance against extinction?

2 Name three causes of heritable variation.

3 Warfarin is a poison used to kill rats. In 1960 a strain of rats appeared which was unaffected by and flourished on this poisonous bait. (a) How did this new strain arise? (b) What will happen in the future to the normal wild type strain of rat if use of warfarin is continued?

4 Arrange the following stages into the correct order in which they would occur during speciation. (a) Increased variation (b) genetic isolation (c) natural selection (d) population explosion (e) physical isolation of gene pools (f) mutation (g) decreased selection pressure.

5

Latin name	Years ago	Cranial capacity (cm^3)
Australopithecus africanus	2 500 000	450–550
Homo erectus	700 000	770–1000
Homo sapiens	30 000	1200–1500

(a) From the facts presented above, which human organ appears to have increased greatly in size during the evolution of man from his ape-like ancestors?

(b) What selective advantage has resulted from this development?

39 Evidence for evolution

Fossils

Normally **fossilisation** involves the conversion of the hard parts of the body (bones, teeth, shells etc.) into **rock**. The age of a fossil can be determined by estimating the date of the rock in which it was found. The older the rock, the less radioactivity it emits.

Evolution of the horse

The horse is one of the few organisms for which sufficient fossils exist to allow its evolution to be studied in detail (figure 39.1). This course of evolution can be explained as follows. Fossil plants in the same rock stratum indicate that *Eohippus* lived in marshy well-wooded country which afforded cover from predators. In such an environment spreading toes provided support and non grinding molar teeth suited a diet of soft fruit. Fossil plants in later rock indicate that the marshy wooded country was gradually replaced by a drier type, thus *Merychippus* lived in open prairie with little concealment. Escape from predators now depended on the head being in an elevated position for gaining a good view of the surrounding terrain and on a high turn of speed using hooves (splayed feet would have proved a hindrance). Survival also depended on natural selection favouring the evolution of long molar teeth with surfaces suitable for grinding grass.

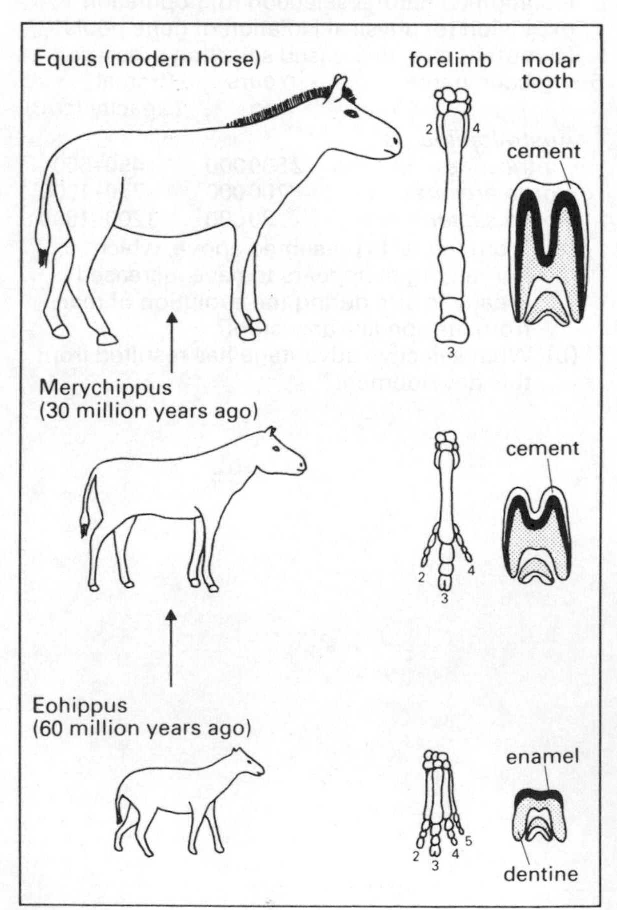

Figure 39.1 Evolution of the horse

The geological timescale

The **geological timescale** (table 39.1) has been formulated from detailed studies of rocks and fossils. Although there are gaps in the fossil record, sufficient fossils exist to suggest that animals and plants have undergone a gradual series of modifications from prehistoric times until the present becoming more and more complex as evolution has progressed.

Homology

Structures are **homologous** if they are structurally alike and have developed from a **common ancestor**. They need not perform the same function.

Pentadactyl (5 digit) limb

This limb type is found in all four classes of terrestrial vertebrates and is traced back to a type of fish closely

Time in millions of years	Geological era	Geological period	animals appearing	plants appearing
	Cenozoic	Quaternary	modern man	
		Tertiary	modern mammals	
100	Mesozoic	Cretaceous		angiosperms
		Jurassic	birds and mammals	
200		Triassic	dinosaurs	gymnosperms
	Palaeozoic	Permian		
300		Carboniferous	reptiles	
		Devonian	amphibians	ferns
400		Silurian		
		Ordovician	fishes	
500		Cambrian	invertebrates	algae
600				

Table 39.1 Geological time scale

related to the ancestors of amphibians. In the course of evolution, the homologous limbs of different vertebrates have become adapted to suit different functions (figure 39.2).

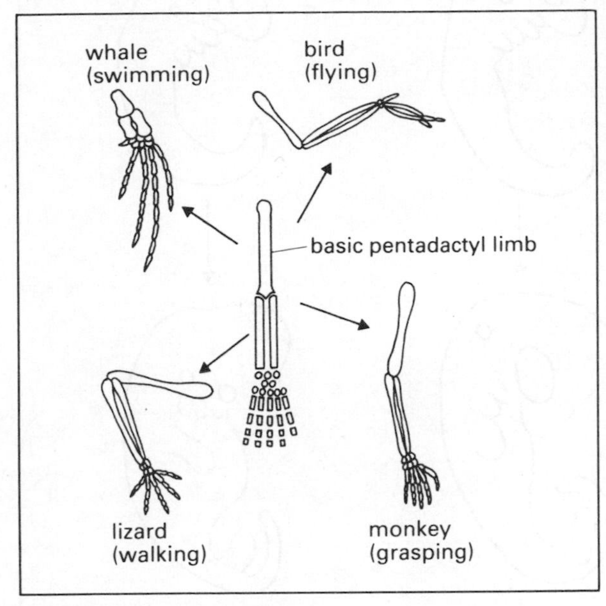

Figure 39.2 Homologous pentadactyl limbs

Vertebrate teeth
Although a lion's canine tooth is sharp and pointed (for stabbing and tearing prey), a sheep's canine small and incisor-like (for cropping grass) and an elephant's canine in the form of a tusk (for defence and offence), they are all similar in basic structure and are therefore homologous.

Pericarp of flowering plants
Each pericarp (ovary wall) shown in figure 39.3 is modified to suit a particular mode of seed dispersal. However they are all similar in basic structure indicating evolution from a common ancestor and again providing evidence for evolution.

Divergent and convergent evolution
When structures are **homologous** but modified to serve different functions, they are said to show **divergent evolution.** When structures are **analogous** i.e. perform the same function but are structurally different (e.g. wings, eyes and legs of a locust and a blackbird) then they are said to show **convergent evolution.**

Embryology

The early **embryonic** stages of all vertebrates are very similar despite the fact that the adult stages are very different (figure 39.4). The more closely

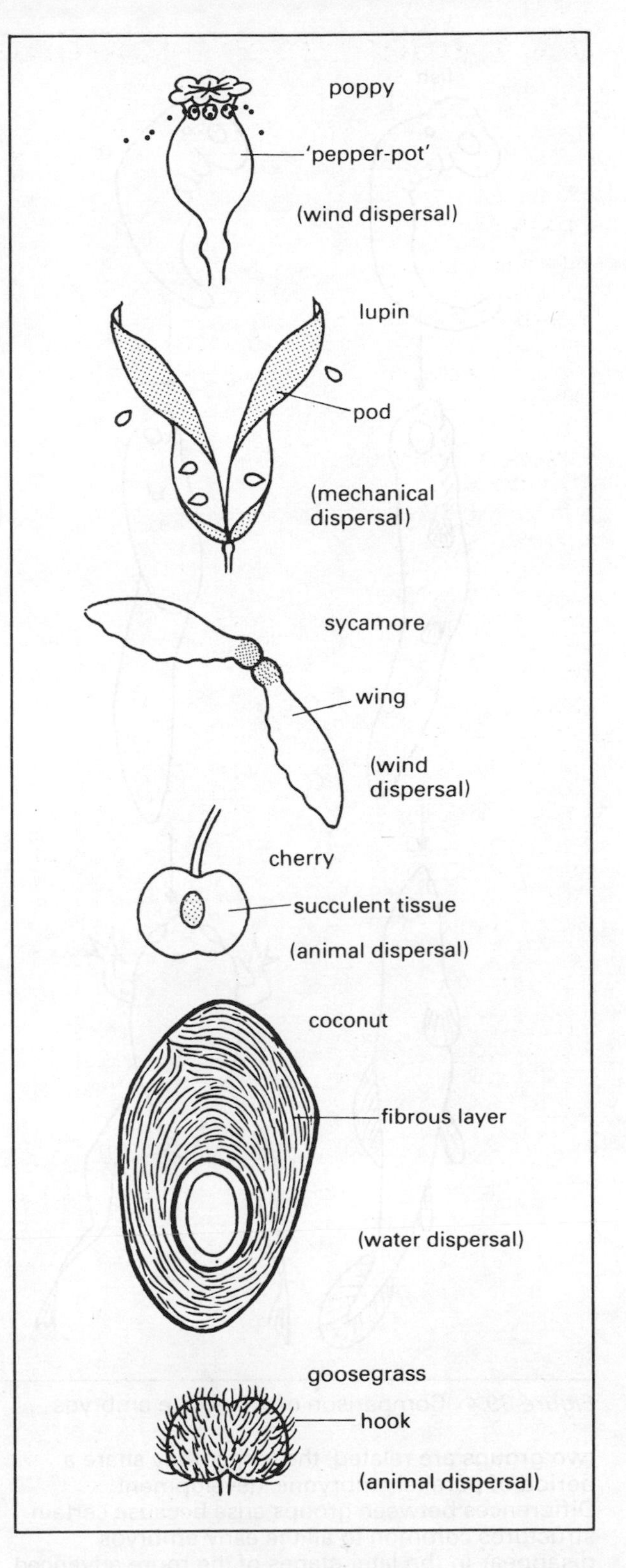

Figure 39.3 Homologous pericarps

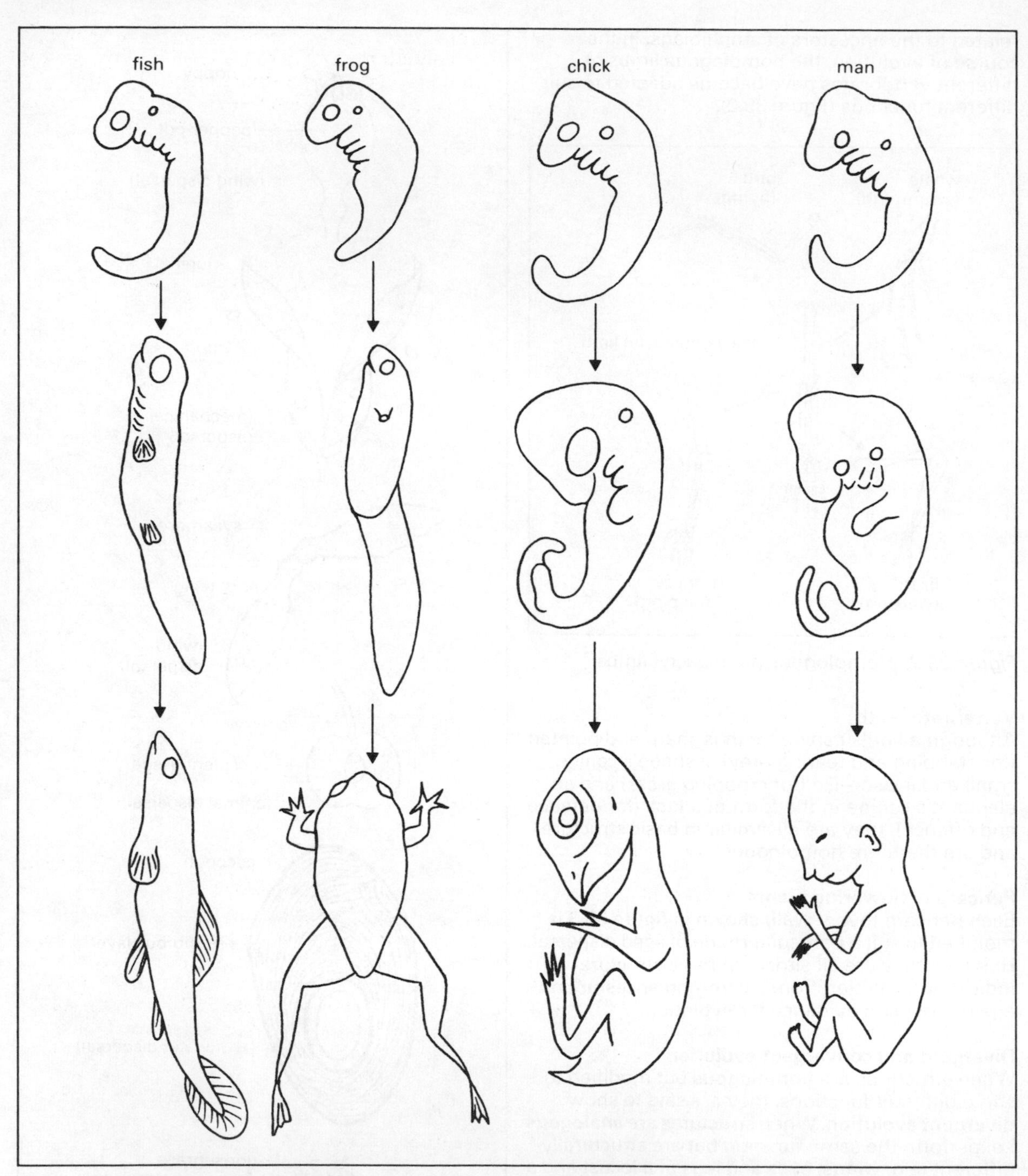

Figure 39.4 Comparison of vertebrate embryos

two groups are related, the longer they share a period of parallel embryonic development. Differences between groups arise because certain structures common to all the early embryos disappear in the later stages of the more advanced ones.

In fish, the embryo has a series of branchial grooves (figure 39.5) matched on the interior by gill pouches. The pouches and grooves eventually meet to form gill slits which allow water to pass out after gaseous exchange. Similarly a frog tadpole has branchial slits when it possesses gills. When the

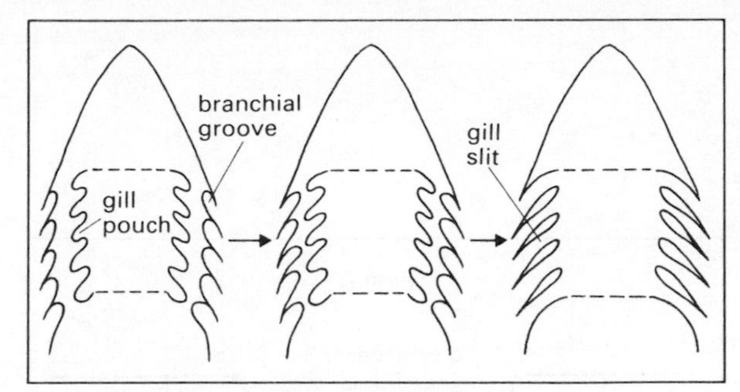

Figure 39.5 Formation of gill slits in a fish embryo

gills degenerate the branchial slits close. In birds and mammals, the grooves and pouches disappear at an early stage and the chief trace of their existence is the Eustachian tube and auditory canal which, interrupted only by the eardrum, connect the throat with the external environment.

The temporary possession of a tail, a two-chambered heart and a single circulation are further examples of developmental stages through which bird and mammalian embryos pass.

It would seem therefore that mammals and other vertebrates continue to pass through many of the embryonic stages that their ancestors passed through because they have all inherited similar **developmental mechanisms** during evolution from a common ancestor.

Geographical distribution

It is thought that ancestral stocks of animals originated in the northern hemisphere and then some of these migrated to southern continents via land bridges. Later they became cut off by continental drift and geographical barriers and therefore each isolated group underwent its own course of evolution. This theory is supported by the fact that the animals native to an area vary greatly from one southern continent to another and are especially different in Australia (presumably first to be isolated). On the other hand the faunas of continents in the northern hemisphere are very similar.

Domestication of plants and animals.

For thousands of years, man has cultivated plants and domesticated animals in order to yield more and better food (see artificial selection, chapter 38). During this process of **selective domestication,** he has produced new varieties of plants and animals which although still the same species, differ greatly from their original ancestors (figure 39.6). The fact that such evolutionary changes (in this case for man's own benefit) can be brought about is further evidence for evolution.

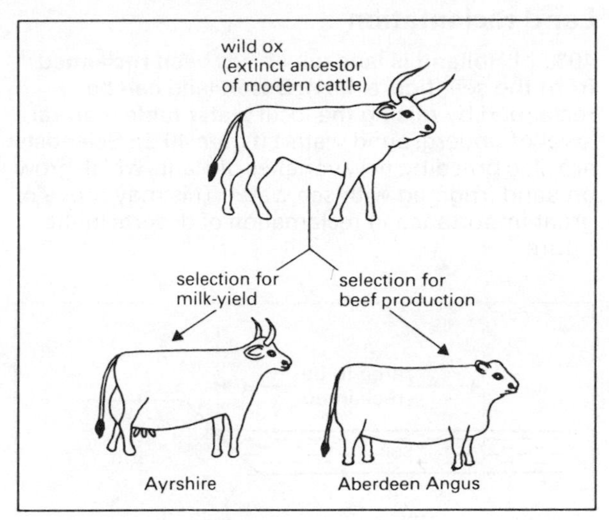

Figure 39.6 Domestication of cattle

Revision questions

1 Why is the evolutionary history of the phylum Mollusca much better known than that of the phylum Annelida?

2 It is known from fossil evidence that there once lived a vertebrate which possessed wings, teeth, feathers and a long tail. Why is this extinct animal regarded as a missing link?

3 An octopus eye is very similar to a vertebrate eye. The former however does not possess an inverted retina and develops in a different manner indicating evolution from an ancestral line unrelated to the vertebrates. (a) Are these eye types homologous or analogous structures? (b) Have they arisen by convergent or divergent evolution?

4 Of the four embryonic stages shown in figure 39.4, which two have most recently shared a common ancestor?

5 New Zealand, at its discovery, was found to possess no mammals except man and bats. Explain.

40 Land use

Land reclamation

40% of Holland is land which has been **reclaimed** from the sea (figure 40.1). Desert land can be reclaimed by raising the local **water table** (natural level of underground water) (figure 40.2). Scientists are also breeding new varieties of plants which grow on sand irrigated with sea water. This may prove of great importance in reclamation of deserts in the future.

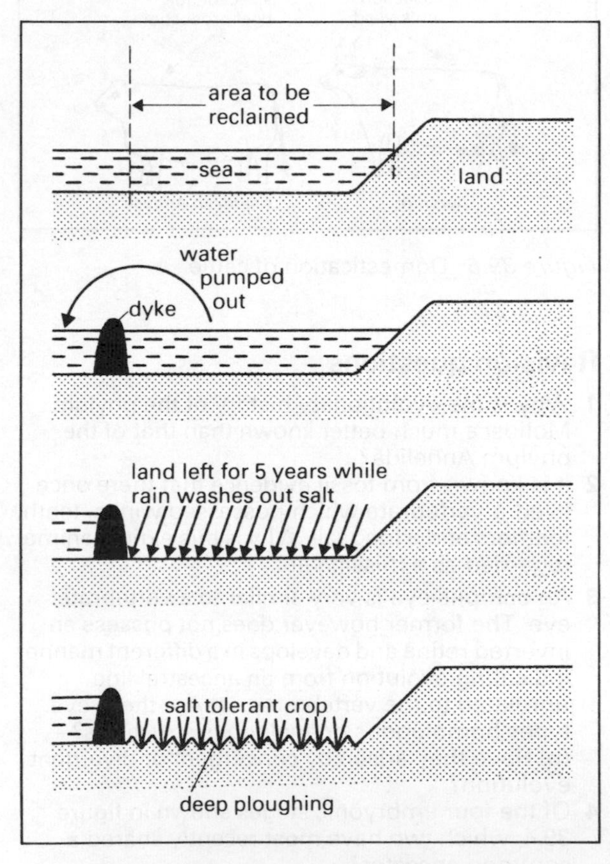

Figure 40.1 Reclamation of land from under sea

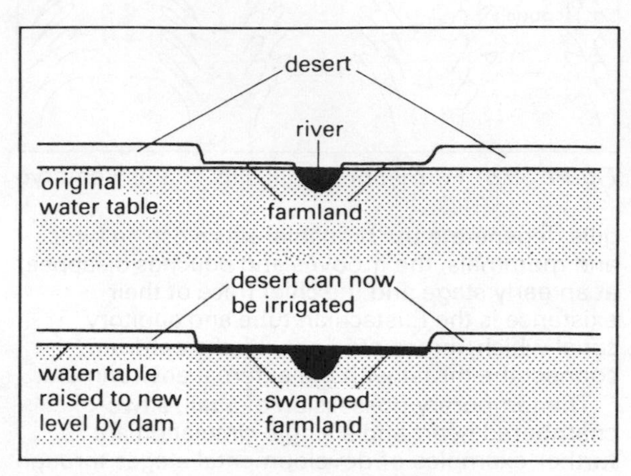

Figure 40.2 Reclamation of desert

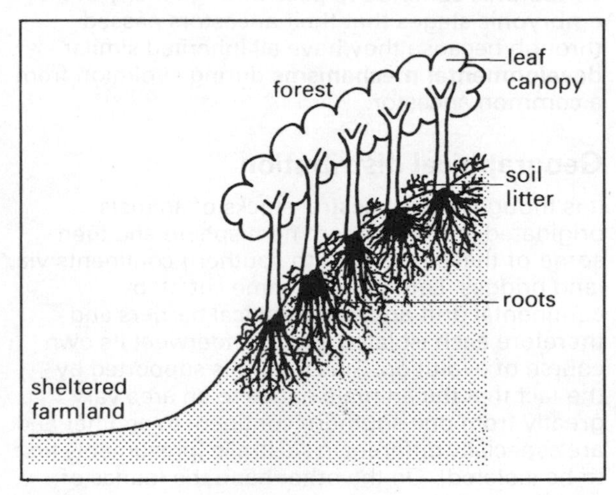

Figure 40.3 An ecosystem free from erosion

Erosion and the effects of roots on the soil

Erosion is the loss of the fertile top soil by the action of wind or water. This rarely occurs in a well forested area (figure 40.3) because the **leaf canopy** reduces the force of falling rain and the **soil litter** acts as a sponge reducing violent sideways flow of water. In addition, roots both open up the soil thus increasing its **porosity** and **bind** the soil particles together preventing erosion.

Afforestation

The conversion of land to forest by planting young trees is called **afforestation**. Because of speed of maturity and suitability to Scottish climate, most trees used are **conifers.** In the past all of the trees in a stand were felled at the same time. However this left large areas of bare soil susceptible to erosion. In addition, neighbouring farmland was deprived of protective cover. Now **block cutting** (figure 40.4) is employed to overcome these problems and still ensure a supply of timber. Afforestation also improves the soil but uses land which could be used for sheep farming.

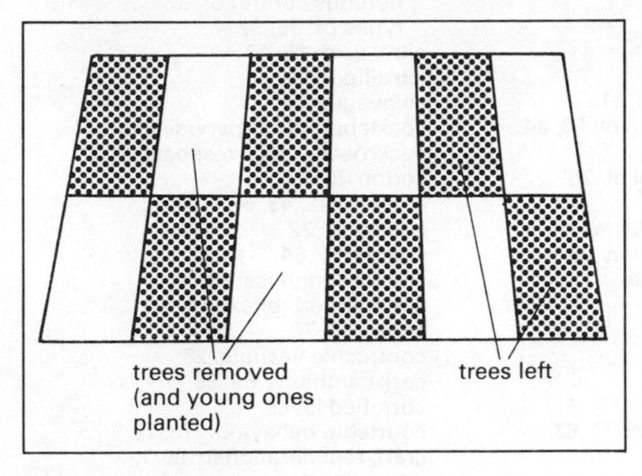

Figure 40.4 Block cutting

Land drainage

Good **drainage** removes excess rain water from the root zone of plants. This ensures **easier cultivation** of land (since it is not so heavy), **faster warming** of the soil in spring, **better germination** of seeds and **deeper rooting** of crops (making them more drought resistant).

Major changes in land use

As a result of the enormous increase in human population and the vast technological advances that have occurred over the past hundred years, the pressures on the 25% of the Earth that is land are increasing daily. In addition to the ever increasing demands on the land to produce more and more food, major changes in land use are also occurring. Land which in earlier times was rural is now rapidly disappearing to be replaced by motorways, factories, housing estates, nuclear power stations, airports and urbanisation in general. To ensure man's successful survival in the future, the most efficient and careful use must be made of the existing land.

Revision questions

1 Predict two results of clearing the forest shown in figure 40.3.
2 Consider the diagram of marram grass shown in figure 40.5. What effect will planting such grass have on sand dunes subject to extremes of wind and rain?
3 Name one advantage and one disadvantage of afforestation.
4 State three characteristics of a poorly drained soil.
5 Name two types of natural area from which land can be reclaimed and made fertile.

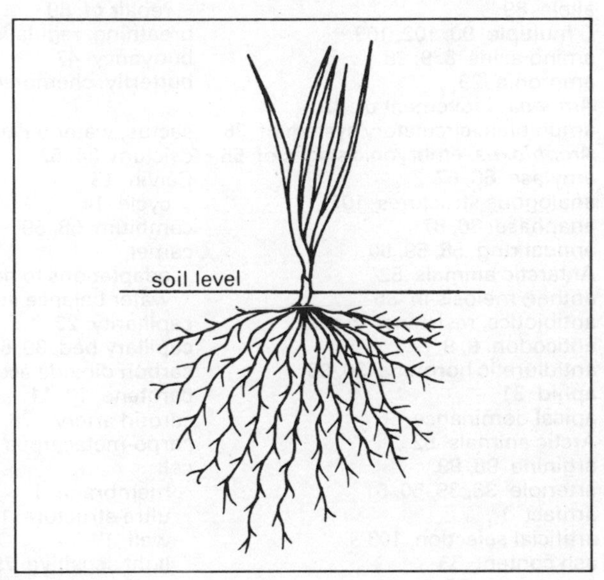

Figure 40.5

Index